Nanostructure Science and Technology

For other titles published in this series, go to
www.springer.com/series/6331

Nanostructure Science and Technology

Series Editor: David J. Lockwood, FRSC
National Research Council of Canada
Ottawa, Ontario, Canada

For other volumes:
http://www.springer.com/series/6331

Vladimir Murashov · John Howard
Editors

Nanotechnology Standards

Springer

Editors
Vladimir Murashov, Ph.D
NIOSH
Washington, DC
USA
vmurashov@cdc.gov

John Howard, MD
NIOSH
Washington, DC
USA
jhoward1@cdc.gov

ISBN 978-1-4614-2800-8 ISBN 978-1-4419-7853-0 (eBook)
DOI 10.1007/978-1-4419-7853-0
Springer New York Dordrecht Heidelberg London

Softcover reprint of the hardcover 1st edition 2011

Printed on acid-free paper

Springer is part of Springer Science+Business Media (www.springer.com)

Preface

Globalization has unleashed economic forces that are affecting knowledge generation, commercial trade in goods and services, and the manufacturing of products. Global economic forces are also leading to a greater role in both commerce and science for international standards. Increasingly, standards are serving an important role in promoting the international development and commercialization of emerging technologies. Standards aid economic globalization by providing a common means to define technical nomenclature, standardize analytical methods, determine whether harmful exposures exist, and provide for ways to control many of the risks associated with international technology commercialization. Also, the development of standards in the twenty-first century to control risks to workers, consumers and the environment is becoming as pivotal to the success of globalization as free trade agreements were in the twentieth century. And, the use of standards in the governance of risk has only increased since nanotechnology has emerged as a global technology which promises to reshape the way we live and work.

Nanotechnology is a rapidly evolving and potentially transformative technology, which has the potential to greatly improve many areas of human life. Nanotechnology promises stronger and lighter materials, more efficacious pharmaceuticals, novel energy sources, more nutritious and longer-lasting foods, more sophisticated national security equipment, and revolutionary cancer treatments. As potentially transformative as nanotechnology may be, however, successful acceptance of any new technology, and its widespread commercial dissemination, requires strict attention to controlling potential risks, especially in countries with robust product liability and personal injury systems. International standards can serve to protect both product users and product manufacturers.

Historically, international standards that have been incorporated into international trade agreements or adopted into national laws have been developed by only a limited number of public and private organizations. For instance, the Organization for Economic Cooperation and Development (OECD), various United Nations organizations, and a number of private organizations, such as the International Organization for Standardization (ISO) and the International Electro-technical Commission have served throughout the twentieth century as the primary route for the development of international standards through a formal national membership requirement. In addition, there also exist a large number of voluntary international standards which are often developed by private organizations without national body memberships, such as ASTM International and the Institute of Electrical and Electronics Engineers.

Existing standards developing organizations or SDOs, both public and private, organize their work through groups of experts focused on specific application areas. With nanotechnology, however, technical groups spanning the entire technology, or very broad aspects of the technology such as environmental safety and health issues, have been formed to coordinate standard setting activities and to allow for sufficient flexibility to accommodate rapidly evolving knowledge about nanotechnology and its potential risks and benefits.

In the last 5 years, almost all major SDOs established such technical groups. For example, ISO established a technical committee for nanotechnologies, TC 229 in 2005, while OECD established Working Party on Manufactured Nanomaterials in 2006. Many of the existing technical groups working in the field of nanotechnology standards development have a number of projects in parallel. Some are aimed at developing a basic terminology for nanotechnology and nanomaterials, some are working to develop specific measurement techniques for nanomaterials, and others are developing occupational and environmental health and safety guidelines.

Nanotechnology Standards reflects this new way of developing international standards for nanotechnology and is organized around broad application areas similar to existing technical groups in various SDOs. An *Introduction* chapter describes history of standards development process, discusses the roles of different standards development bodies active in nanotechnology, outlines the context of national and international standards development for nanotechnology, highlights the use of knowledge management systems in twenty-first century standards development, and discusses the unique challenges of "proactive" standards development, such as how to reach consensus under the conditions of limited knowledge. Next, there are chapters providing state-of-the-art reviews on developments in topical areas of *Nomenclature & Terminology; Reference Materials; Metrology; Performance Standards; Application Measurements; Implication Measurements; Biological Activity Testing; and Health and Safety*. Each of these chapters summarizes the active areas of national and international standards development and describes the knowledge base to support current nanotechnology standards and future directions in nanotechnology standards development. Finally, the chapter on *Legal Considerations* puts standards development in the context of international legal requirements and application of international standards to national governance structures.

Nanotechnology Standards is the first comprehensive collection of state-of-the-art reviews of twenty-first century nanotechnology standards development written by an international team of experts representing both the international SDO community and the nanosciences community. The authors reflect a diversity of intellectual views and global geographies. The book captures the most recent developments and outlines future directions in the dynamic field of international and national nanotechnology standards development. This book is an essential reference for a broad range of nanotechnology and materials scientists, engineers, lawyers, regulators and students in academic, industrial and government settings who are dealing directly with developing nanotechnology products or with managing the risks of nanotechnology or who just want to learn more about how to manage such risks using nanotechnology standards.

Washington, DC

Vladimir Murashov
John Howard

Contents

Contributors

Shuji Abe
National Institute of Advanced Industrial Science and Technology,
Tsukuba, Japan
s.abe@aist.go.jp

Chris Bell
Sidley Austin LLP, Washington, DC, USA
cbell@sidley.com

Werner Bergholz
Jacobs University, Bremen, Germany
w.bergholz@jacobs-university.de

Jennifer E. Decker
Institute for National Measurement Standards,
National Research Council of Canada, Building M-36, 1200 Montreal Road,
Ottawa, ON K1A 0R6, Canada
Jennifer.Decker@nrc-cnrc.gc.ca

Hendrik Emons
Institute for Reference Materials and Measurements,
Joint Research Centre of the European Commission, Geel, Belgium
Hendrik.EMONS@ec.europa.eu

David S. Ensor
RTI International, Research Triangle Park,
Durham, NC, USA
dse@rti.org

Norbert Fabricius
Karlsruhe Institute of Technology, Karlsruhe, Germany
norbert.fabricius@kit.edu

John Howard
National Institute for Occupational Safety and Health, Centers for Disease
Control and Prevention, U.S. Department of Health and Human Services,
Washington, DC, USA
JHoward1@cdc.gov

Shingo Ichimura
National Institute of Advanced Industrial Science and Technology (AIST),
1-1-1, Umezono, Tsukuba 305-8568, Japan
s.ichimura@aist.go.jp

Fred Klaessig
Pennsylvania Bio Nano Systems, LLC, Doylestown, PA, USA
fred.klaessig@verizon.net

Laurie E. Locascio
Biochemical Science Division, National Institute of Standards and Technology,
Gaithersburg, MD 20899-8310, USA
locascio@nist.gov

Martha Marrapese
Keller and Heckman, LLP, Washington, DC, USA
marrapese@khlaw.com

Vladimir Murashov
National Institute for Occupational Safety and Health, Centers for Disease
Control and Prevention, U.S. Department of Health and Human Services,
Washington, DC, USA
vmurashov@cdc.gov

Hidehiko Nonaka
National Institute of Advanced Industrial Science and Technology (AIST),
1-1-1, Umezono Tsukuba 305-8568, Japan
hide.nonaka@aist.go.jp

Richard C. Pleus
Intertox Inc., Seattle, WA 98101, USA
rcpleus@intertox.com

Georg Reiners
Federal Institute for Materials Research & Testing, Berlin, Germany
georg.reiners@bam.de

Vytas Reipa
Biochemical Science Division, National Institute of Standards
and Technology, Gaithersburg, MD 20899-8310, USA
vytautas.reipa@nist.gov

Gert Roebben
Institute for Reference Materials and Measurements,
Joint Research Centre of the European Commission, Geel, Belgium
gert.roebben@ec.europa.eu

Alan G. Steele
Institute for National Measurement Standards, National Research Council of Canada, Building M-36, 1200 Montreal Road, Ottawa, ON K1A 0R6, Canada
alan.steele@nrc.cnrc.gc.ca

Justin M. Zook
Biochemical Science Division, National Institute of Standards and Technology, Gaithersburg, MD 20899-8310, USA
justin.zook@nist.com

Chapter 1
Introduction*

Vladimir Murashov and John Howard

1.1 Introduction

A standard can be understood as a rule, norm or requirement that is broadly established chiefly by authority, custom or consent. The American National Standards Institute (ANSI) classifies standards by function or origin into eight types: basic, product, design, process, specification, code, management system and personnel certification standard [1]. Historically, standards were developed in limited geographies in parallel with man's own technical development by common use and early custom.

Now, international standards are mostly developed by organized groups of stakeholders assembled from around the world and focus primarily on facilitating communication, promoting commercial trade and ensuring safety and health. Worldwide, there well may be more than 500,000 standards developed by more than 1,000 standard-setting bodies [1]. The number of standards, as well as the geographic and technical boundaries of their application, have been increasing as knowledge about societal risks, and the rapid communication of that knowledge, have grown. Standards development and standards application in governance have also been evolving to reflect changes in world trade, transportation, economics and politics. This chapter discusses the history of standards development in general, and the emerging area of standards development for nanotechnology in particular.

---

*The findings and conclusions in this report are those of the authors and do not necessarily represent the views of the National Institute for Occupational Safety and Health.

V. Murashov (✉)
National Institute for Occupational Safety and Health, Centers for Disease Control and Prevention, U.S. Department of Health and Human Services,
Washington, DC, USA
e-mail: vmurashov@cdc.gov

V. Murashov and J. Howard (eds.), *Nanotechnology Standards*,
Nanostructure Science and Technology, DOI 10.1007/978-1-4419-7853-0_1,

1.2 History of Standards

The history of standards development can be broken into four phases, each characterized by distinct types of standards development, distribution modalities and governance mechanisms: (1) community standards development; (2) national standards development; (3) international standards development; (4) global standards development.

1.2.1 Community Standards Development

The history of standards development can arguably be traced to the first use of time-unit standards by ice-age hunters in Europe over 20,000 years ago. These early standards developers scratched lines and gouged holes in sticks and bones for the purpose of counting the days between phases of the moon [2]. The first standards were aimed at harmonizing human activities with natural phenomena. The functions and applications of standards have been expanding ever since and have followed closely the increasing complexity of man's own technical and societal development.

The advent of agriculture as early as 10,000 years ago was a critical step in the development of human civilization and in the course of standards development. The production of excess food, made possible by agriculture, created a foundation for trade. In turn, trade required the introduction of unit standards to ensure that trade was fair, and to collect taxes on the goods that were traded. These early standards, such as unit standards for product value (or money) [3], length, and weights [4], were enforced by local authority or, in some cases, the state. Examples of commercial trading standards, such as measurement and exchange of goods can be found in the Babylonian Code of Hammurabi created in 1790 BCE.

Agriculture also facilitated the development and accumulation of technical knowledge and the standards that were associated with that knowledge. The need to transfer accumulated knowledge effectively to subsequent generations in the form of standards led to the development of writing. The first preserved writing was created 5,000 years ago in Egypt, Mesopotamia and China [5]. Writing was used to convey and distribute sophisticated standard practices for every activity. For example, the Horus Temple in Edfu, Egypt contains second century BCE carvings of practice standards specifying formulas for preparing incense and ointment for the divine statues [6].

Practice standards aimed at safety and health, such as building codes specifying the minimum acceptable level of safety for constructed objects and structures, appeared initially in agricultural societies. One of the first preserved building codes can be found in the Code of Hammurabi. Early food safety standards, another example of safety and health practice standards, were established primarily to prevent economic deception and adulteration of foods [7]. For example, the Romans wrote civil law provisions to protect the populace against adulterated foods. In 200 BCE, the Roman statesman Cato described a method for determining whether merchants "watered down" their wine [7]. The English passed their first food law, the Assize of Bread, in 1266 to prevent the adulteration of bread with cheaper, inferior ingredients. The German "beer purity" law ("*Reinheitsgebot*") of 1516 was the oldest existing food safety regulatory standard in the world until it was struck down as a trade barrier by the European Court in 1987. The *Reinheitsgobot* gave the government the tools to regulate the ingredients (limiting them to barley malt, hops, and water), the processes and the quality of beer sold to the public [8]. In early colonial America, early food regulations were aimed at promoting export of quality food to Europe. For example, the Massachusetts Bay Colony's Meat and Fish Inspection Law of 1641 was developed to demonstrate that the colony produced and exported high-quality food products to the mother country, thereby gaining commercial advantage [7].

Artisanship and crafts flourished in agricultural societies. "Secrets of the craft" – a type of proprietary practice standard – was used by craftsmen and artisans from a wide range of trades such as masonry, glasswork and carpentry. These standards evolved with the sophistication of the particular craft and formed the basis of medieval guilds. Such practice standards, along with rules of professional conduct, were developed and practiced by craftsmen associations and passed from master to apprentice through many generations. Such practice standards were developed as early as 200 BCE in China where guilds, known as "*hanghui*," existed during the Han Dynasty. These guilds survived through the centuries and still exist in China for certain professions [9]. Gaining and protecting a particular guild's competitive advantage over less skilled market entrants was a prime function of such associations. Competitive advantage was facilitated by restricting knowledge of the practice standards to members of the association only. By the eighteenth century, these associations became obstacles to free trade and hindered technological innovation, technology transfer and business development [10]. As a result, they were replaced with national trade associations which developed transparent standards for all to see and use.

1.2.2 National Standards Development

The advent of the steam engine and the industrial revolution of the mid-nineteenth century facilitated the emergence of powerful national states and decentralization of manufacturing and trade throughout those states. This created the need, for example, for nationally harmonized specification standards for transportation, such as a standard railroad gauge [11], and material specification standards, such as grades of steel used in rail construction [12]. These needs were addressed through voluntary standards development by national standards developing organizations (SDOs) and trades associations. Examples of such national standards development associations include the American Society for Testing and Materials (ASTM) and the Institute of Electrical and Electronics Engineers (IEEE). ASTM, which is presently known as ASTM International, was formed in 1898 in the United States of America (USA) by a group of engineers and scientists to address frequent rail breaks [13]. This ASTM work led to standardization of the steel used in rail construction across the USA. The IEEE, the world's largest technical professional society, was formed as the American Institute of Electrical Engineers in 1884 to support electrical professionals [14]. Both ASTM and IEEE later evolved to become private international consensus standards developing organizations without national body membership. Well over 600 SDOs currently exist in the US, some quite small with few standards while others are global in every sense.

Many national standards developing organizations, including those in United Kingdom (UK), in the USA and in the Russian Federation (Russia), were established at the end of the nineteenth century not only to harmonize and manage nationally-developed standards across nations, but also to represent national interests in international standards developing organizations.

The origin of the UK National Standardization Body (presently known as BSI British Standards) can be traced to the Engineering Standards Committee (ESC) which was founded by the Council of the Institution of Civil Engineers in 1901 [15]. The ESC extended its work to other fields, and was renamed to the British Standards Institution in 1931 after receiving its Royal Charter in 1929 [15].

The predecessor to ANSI, an administrator and coordinator of the United States private sector voluntary standardization system, was formed in 1916. ANSI was created when the American Institute of Electrical Engineers invited the American Society of Mechanical Engineers, the American Society of Civil Engineers, the American Institute of Mining and Metallurgical Engineers, and the American Society for Testing and Materials, to join together to establish a neutral national body to coordinate standards development, approve national consensus standards, and prevent user confusion on acceptability criteria [1].

In Russia, the Federal Agency on Technical Regulation and Metrology ("*Rostekhregulirovaniye*") serves as the National Standardization Body. It was formed by a Russian Presidential decree as a successor to the Union of Soviet Socialist Republics' State Standards agency upon disintegration of the USSR in 1991. In turn, the USSR State Standards agency originates from the Committee on Standardization established in the Soviet Council on Labor and Defense in 1925 [16].

Models for national standards development vary from country to country. In the USA, there is no official National Standardization Body in the sense of the UK and Russia as described above. Instead, standards development is voluntary, and overwhelmingly, private sector led. The National Technology Transfer and Advancement Act of 1995 (Section 12(d) of P.L. 104-113), and OMB Circular No. A-119 [17] directs Federal government agencies to use voluntary consensus standards in lieu of government-unique standards except where inconsistent with law or where otherwise impractical to implement. In addition, these documents also encourage Federal government agencies to actively participate in the development of voluntary consensus standards. For example, Occupational Safety and Health Administration (OSHA), a USA government regulator of occupational safety and health, and ANSI, have signed a Memorandum of Understanding which states in part that "ANSI will furnish assistance and support and continue to encourage the development of national consensus standards for occupational safety and health issues for the use of OSHA and others" [18].

The USA government permits participation in voluntary consensus standards development by government personnel, but does not provide dedicated funding to do so. For example, Public Law 107-101, National Defense Authorization Act for Fiscal Year 2002 (Public Law 107-101, Section 1115, entitled "Participation of Personnel in Technical Standards Development Activities") nullified language in USA Code Section 5946 which restricts the use of appropriated funds for payment of membership dues or expenses of an individual at meetings or conventions of members of a society or association. Yet, even though statutory permission exists for government experts to participate in voluntary standards development, government appropriations often do not provide the funding necessary for a government agency to send their experts to meetings. In addition, academic researchers receiving government grants and government personnel receiving salary and research support often do not have specific funding to participate in standards development committees. This approach is not supportive of the ANSI funding model, which is based on membership fees collected from volunteers participating in the development of standards. Nevertheless, as of January 2008, ANSI lists 267 entities accredited by ANSI as standards developers including: entities suggested by government regulators; entities prompted by industry; entities open to individual international membership, but without any established one-vote-per country membership structure; and entities with either long or short term missions. There is also a host of other US-based standards developers operating outside the ANSI accreditation, including some that produce standards in global use.

At the other end of the spectrum are countries where standards development is completely a government function. In Russia, the Federal Agency on Technical Regulation and Metrology ("*Rostekhregulirovaniye*") provides funds for the development of *GOsudarstvennyi STandart-Rossii* (GOST-R) standards, which are the national standards of the Russian Federation. The Russian Federal Agency on Technical Regulation and Metrology also administers a GOST-R certification program for products and acts as a national member body in international standards organizations such as International Organization for Standardization (ISO). Similarly, the Standardization Administration of China (SAC) is authorized by the State Council to exercise administrative responsibilities for standardization work in China and serves as a national standardization body in international standards organizations, such as ISO [19]. SAC is responsible for funding and managing the development of national standards including research in support of national standards development and maintenance.

Standards development in the UK represents an intermediate case with the standards development enjoying some level of government support. The UK National Standards Body, British Standards Institution (BSI), can act on behalf of the UK government in some instances [20]. It is a non-profit organization which markets for sale the various standards it develops. BSI British Standards develops standards through committees including representatives from government, testing laboratories, suppliers, customers, academic institutions, business, manufacturers, regulators, consumers and trade unions. It also provides testing and certification services.

Regardless of how national standards developing organizations function, they have facilitated the emergence of standards which provide consistency across divergent commercial entities operating within national boundaries. But, with the increasing volume of international trade, differing national standards have created significant obstacles to global trade [21]. The drive to harmonize national standards between countries has prompted establishment in the early twentieth century of international standards developing organizations with national body membership.

1.2.3 International Standards Development

International standards often provide technical foundation for international agreements or treaties. They are based on the voluntary participation of all national commercial market interests which are affected by the particular standard. Establishment of international standards, such as international codes of conduct, was catalyzed by international trade in the late nineteenth and early twentieth centuries. Early international governance standards can be traced to the emergence of seafaring as a valuable way to trade goods.

The earliest known examples of sailing instructions – the Greek *periploi*, which became the basis for cartography standards in the medieval period – date from the fourth to third centuries BCE [22]. Modern practice standards governing conduct at sea originate from a 1609 work of Grotius, a Dutch lawyer, titled *Mare Liberum* ("Free Seas"). Grotius articulated the principle of the "freedom of the seas." This principle held that the sea is not owned by any particular nation, but should be available for use by all nations. In 1884, another milestone in establishing maritime standards occurred when the International Meridian Conference, a public government-level organization, adopted the Greenwich meridian as the universal prime meridian or zero point of longitude, undoubtedly the historically most important example of a maritime reference standard.

The twentieth century achievements in commercial aviation, and the post-WWII rapid economic growth spurt, further accelerated the development of international standards. As trade and transport between countries increased, so did the need for international standards and international standards developing organizations. Specifically, multinational enterprises and other transnational actors sought to harmonize national legal standards applicable to international transactions (such as safety and health testing standards) to reduce the transaction costs of doing business across nations [23].

A wide range of international standards developing organizations emerged in mid-twentieth century targeting specific gaps in standards. International standards incorporated into national laws and international agreements are often developed by only a limited number of public organizations with national body memberships such as the Organization for Economic Cooperation and Development (OECD), the International Labour Office (ILO) and the World Health Organization (WHO), and a number of private organizations such as the International Organization for Standardization (ISO) and the International Electro-technical Commission (IEC). However, there also exists a large body of voluntary international standards which are often developed by private organizations without national body memberships, such as ASTM International and IEEE. Nevertheless these standards can be globally accepted and used by industrial interests. With the wide range of standards development models, the governance model where a public standards developing organization identifies essential requirements, considers voluntary standards which were developed by private standards developing organizations, and sets technical specifications to meet essential requirements, has become more prominent in the late twentieth century and early twenty-first century [23].

Inter-governmental organizations have been involved in international standards development since the League of Nations was established in 1919 under the Treaty of Versailles "to promote international cooperation and to achieve peace and security" [24].

Founded in 1945, the United Nations (UN) can be considered the largest public international organization developing standards. The UN Charter contains a broad mandate "to achieve international co-operation in solving international problems of an economic, social, cultural, or humanitarian character" and "to be a center for harmonizing the actions of nations in the attainment of these common ends" [25]. Some UN specialized agencies were even established before 1945 and subsequently incorporated into the UN. For example, the International Telecommunication Union was founded in 1865 as the International Telegraph Union to facilitate communication between nations, and the ILO was founded in 1919 to promote social justice and international human and labor rights.

The General Assembly occupies a central position as the UN's chief deliberative, policymaking, and representative body. It comprises all 192 member nation and plays a significant role in the process of standards development and the codification of international law [25]. The General Assembly is empowered to make non-binding recommendations to member nations on international issues within its competence. Each member nation in the Assembly has one vote. Votes taken on designated important issues, such as recommendations on peace and security and the election of Security Council members, require a two-thirds majority of member nation, but other questions are decided by simple majority. In recent years, a special effort has been made to achieve consensus on issues, rather than deciding by a formal vote, thus strengthening support for the General Assembly's decisions.

Technical standards in UN are commonly developed at committee or task group levels by experts nominated by participating member nations. The development process varies for different standards areas covered by UN organizations. For example, the Codex Alimentarius Commission was created in 1963 by Food and Agriculture Organization (FAO) and WHO to develop food standards including guidelines and related texts such as codes of practice under the Joint FAO/WHO Food Standards Programme [26]. The main purposes of this Programme are protecting health of the consumers, ensuring fair trade practices in the food trade, and promoting coordination of all food standards work undertaken by international governmental and non-governmental organizations. Decisions are taken by a majority of the votes cast at annual meetings of the Commission with each member nation of the Commission having one vote [27].

Another example is standards development process in the World Health Organization. The WHO functions according to its constitution include development of health-related standards [28]. Although major regulatory standards are adopted at the World Health Assemblies by member nations, in the development of most technical standards WHO historically has relied on expert opinions obtained through Expert Advisory Panels and Committees [29]. These panels are convened to make technical recommendations on a subject of interest to WHO. Advisory panel members are appointed by the WHO Director General and these members contribute technical information and offer advice on scientific developments in the expert's field. However, the WHO process has been criticized as promoting poor quality standards. In 2003, WHO improved the process with the publication of WHO Cabinet Guidance [30, 31]. WHO emphasized the use of an evidence-based and transparent approach to the development of standards and implementing standards [32]. Once the draft standards are prepared, they are approved in most cases by the Director-General Office [28].

In contrast, ILO is the only tripartite UN agency. To carry out its responsibilities for drawing up and overseeing international labor standards, ILO gathers representatives of government, employers and workers to develop its policies and conventions jointly.

The OECD is a treaty organization which can be also considered a public sector transparent international standards developing organization. OECD is both a user of international standards and a developer of standards (technical regulations) to address needs of OECD member governments. OECD was formed as the Organization for European Economic Co-operation in 1947 to administer US and Canadian post-World War II aid under a specific reconstruction plan [33]. In 1961, the Organization for European Economic Co-operation became OECD with a mission to help member countries achieve sustainable economic growth, robust employment, and high standard of living. Today, OECD is composed of 34 member countries committed to democracy and a market economy. OECD shares its member countries' expertise with more than 70 other countries. In addition, OECD invited Russia to engage in membership talks and offered enhanced engagement to Brazil, China, India, Indonesia and South Africa. Industry and labor have also been engaged with the OECD since its creation, notably through the Business and Industry Advisory Committee to the OECD and the Trade Union Advisory Committee to the OECD.

OECD has established an effective standards development process consisting of data collection, data analysis, and collective policy discussions, followed by collaborative decisions-making and implementation. Technical work in OECD is conducted through committees and working parties by representatives of member countries and by invited non-member experts. Discussions at the OECD committee level can culminate in formal agreements with countries to produce specific standards or model recommendations or guidelines such as Good Laboratory Practices [34]. According to the Convention on the OECD, which established a legal framework for OECD operations, decisions made by the OECD are binding on all OECD Members (Article 5(a), [35]). Decisions and recommendations are made by consensus defined as "mutual agreement of all the Members" (Article 6(1), [35]) with each member holding one vote during the adoption process.

The first private sector international standards developing organizations with national body memberships, the International Electro-technical Commission, held its inaugural meeting in June 1906, following the recommendation of the 1904 International Electrical Congress. IEC prepares and publishes international standards for electrical, electronic and related technologies and manages conformity assessment systems. In 2008, the IEC listed 72 members and 83 affiliate country members developing standards by means of 174 technical committees and subcommittees.

The ISO is perhaps the most well-known private sector international standards developing organizations with national body memberships. The ISO arose out of the International Federation of the National Standardizing Associations (ISA) which was established in New York City in 1926. The ISA focused heavily on mechanical engineering and was disbanded in 1942. In 1944, the UN established the United Nations Standards Coordinating Committee (UNSCC). In 1946, ISA was re-established and then was merged with UNSCC in 1947 to create ISO [36, 37]. Today, the ISO has a membership of 160 national standards institutes with about 680 international standards developing organizations as partners, including most UN agencies.

In ISO and IEC, each national member body has one vote as in the UN. Standards are based on two levels of consensus: (1) consensus between national stakeholders to put forward as a national position and (2) consensus across nations. As a reflection of the diverse nature of the support from national standards development bodies, some ISO and IEC national members are either part of the governmental structure in their countries or have a mandate from their governments to engage in international standardization, while others are private sector standards development bodies originating in industrial associations. Participation of developing countries in ISO and IEC is facilitated by pro-rating membership fees on the basis of the each nation's Gross Domestic Product.

Two major private international standards developing organizations without national body memberships are ASTM International, and IEEE Standards Association. Both follow the principles of transparency and consensus in developing international standards similar to ISO and IEC except for the "one-country-one-vote" principle. Membership in these organizations is open to any individual, company, governmental agency, academia, or similar entity, upon payment of annual fees. Every individual member or entity has one vote. Therefore, both organizations benefit from international membership from a wide spectrum of technical experts. For example, in ASTM International, a diverse range of standards is developed by over 30,000 members, representing producers, users, consumers, government and academia from over 120 countries [38]. IEEE Standards Association has over 20,000 members participating in standards development for electro- and information technologies and sciences [39].

The main objective of developing consensus-based international standards through diverse public and private organizations is to facilitate global trade and to protect human health and the environment. As such, they are widely used to support the regulatory work of global intergovernmental organizations such as the World Trade Organization (WTO) and the OECD.

Specifically, when it comes to international trade, the WTO Agreement on Technical Barriers to Trade explicitly recognizes the importance of international consensus standards. For instance, an importing nation's requirement that imported goods trade must conform to an international standard does not constitute a basis under WTO rules for an exporting nation to claim that the importing country is erecting a "trade barrier" [40]. Similarly, at the OECD Ministerial meeting in 1997, the role of international standards received a big boost from the policy recommendation to "develop and use, wherever possible, internationally harmonized standards as a basis for domestic regulations, while collaborating with other countries to review and improve international standards to assure that they continue to achieve intended policy goals efficiently and effectively" [41].

Economic forces unleashed by globalization in knowledge generation, trade, manufacturing and safety oversight have led to promoting the role of international standards development. As twenty-first century international politics move towards more globally distributed technological power, the worldwide acceptance of US national standards produced outside the more formal international standards framework as de facto international standards – common in the twentieth century when the US economy dominated the world – has been diminishing [42].

In the early phase of twenty-first century standards development, international organizations are providing increased opportunities for negotiations between representatives of divergent national economic interests. This process, though time-consuming, has worked well in the twentieth century. However, the revolution in communication created by new digital information technologies which have expanded social networks to a global scale and facilitated further globalization of commerce and production has put new pressures on traditional international standards developing organizations and created new opportunities for emergence of global standards developing frameworks.

1.2.4 Global Standards Development

The emergence of the global phase of standards development is marked by a revolution in information technology which is leading to a radical shift in the standards development process.

On September 2, 1969 engineers at the University of California at Los Angeles transferred data from one computer to another, which signified the beginning of the internet [43]. By the 1990s, the World Wide Web brought the internet from the academic environment to mainstream users. Instant access to the data, information and knowledge, real-time exchange of ideas, and creation of documents by a specialized groups of experts located in different countries, became not only possible, but commonplace. "Cloud" computing and "data farming" are revolutionizing how new knowledge is generated, analyzed and disseminated [44].

Information technology has also become a tool to facilitate the standards development process in several different ways. In the early twenty-first century, electronic balloting, which increases participation by reducing travel costs, was adopted by most standards developing organizations. Knowledge management systems were introduced to facilitate the connection between the generation of new knowledge and the development of standards dependent for their relevance on new knowledge. This connection was facilitated by the emergence of novel information technology capabilities. Knowledge management systems are revolutionizing the standards developing process by democratizing it and reducing time-lag between knowledge generation and standards adoption. Using the knowledge management approach, websites have been established for the development of consensus-based dynamic global standards. For example, in October 2009, ISO launched the ISO Concept Database to provide an environment for ISO committees to store and develop structured content including terms and definitions, graphical symbols, codes, data dictionaries, product properties, and reference data used in their standards [45]. The database is available to the public and allows the public to obtain terms and definitions, graphical symbols, codes, data dictionaries, product properties, and reference data free of charge without buying the standards containing them.

A "wiki," from the Hawaiian word for "fast," is a website that connects interlinked Web pages to aid in collaboration among geographically separated parties. Wikis are powered by wiki software. A wiki-software platform for the generation and maintenance of consensus documents, including standards, is one novel approach made possible by internet-based knowledge management systems. The most well-known example of using this approach is *Wikipedia*. Wikipedia's predecessor, *Nupedia*, was created in 2000 as a platform for expert-written, peer-reviewed content [46]. However, *Nupedia* failed as it was based on the traditional model for content generation and quality assurance. A new model that did not have a formal editorial review process, www.wikipedia.org, went "live" on the internet in January 2001. In the new model, the quality of the content was assured by volunteer editors who checked their own and others' contributions to content against Wikipedia rules [47]. As the project matured, vandalizing or diluting its content became rare, while the accuracy significantly improved and approached that of the Encyclopedia Britannica [48]. By 2009, Wikipedia has grown to a massive global enterprise containing more than 13 million articles in 271 different languages with a budget of $6 million US per year [46]. The success of web-based platforms did not go unnoticed by the traditional standards developing organizations. In 2008, ANSI started utilizing a "wiki-platform" to facilitate its own standards development processes [49].

Advancements in information technology have also democratized the standards development process by significantly reducing entry and participation costs and have enabled new standards development entities to emerge. In September 2008, a consortium of stakeholders launched the *GoodNanoGuide* project based on a wiki software platform [50]. The *GoodNanoGuide* is described as a "collaboration platform designed to enhance the ability of experts to exchange ideas on how best to handle nanomaterials in an occupational setting. *GoodNanoGuide* meant to be an interactive forum that fills the need for up-to-date information about current good workplace practices, highlighting new practices as they develop" in a fast-moving area of technology [50]. New entities like the *GoodNanoGuide* could prove to be viable alternatives to the traditional standards developing organizations provided that issues of transparency, credibility, funding, and quality assurance can be resolved.

The *wiki*-based model for standards development can be further enhanced through addition of automatic programs for annotation and embedding media files such as sound files, videos and charts and, more importantly, for automatic update as linked data is changed [51]. Such new tools, characterized by real-time authoring, date-stamped recording of contributions, and automatically updated live content, could prove to be a useful method for the development and maintenance of dynamic standards. It would also prompt the further evolution of the process for creating and maintaining global standards.

Progress in information technology has also contributed to the emergence of a global community with access to standards development. At the beginning of the twenty-first century, the influence of the public over national and international safety and health regulation increased significantly. Economic globalization and the involvement of formerly national grassroots interests participating in international standards development has put pressure on national and international standards developing organizations to use the standards development process to guide technological innovation to ensure a safe and healthy outcome for workers, consumers and for the environment. For example, as a reflection of this shift, ISO's 2011–2015 Strategic Plan states that its activities aim to address five global challenges:

1. "Facilitation of global trade in products and services in a way that does not compromise the level of safety and quality of life to which the citizens of the global village aspire, in the context of an overall increasing, but also, in some regions, aging world population;
2. Financial crisis which started in 2008, and which affects financial markets and impacts on economies at large, has shown the need to restore confidence, to promote good business and governance practice, and to better anticipate and manage risk and business continuity;
3. Interrelated challenges of responding to climate change, ensuring a sustainable energy future, optimizing the use of, and access to, water and providing the world's growing population with adequate food supplies in a safe and sustainable way;
4. Pervasiveness and rapid growth of information and communication technologies, which revolutionize daily life as well as production processes and business practice; and
5. UN Millennium Goals of eradicating poverty and hunger and granting access to education and better health conditions to all the people of the world" [52].

Addressing the challenges that ISO points out in its Strategic Plan requires a more "proactive" approach to standards development. The "reactive" approach does not look down the road to see future stumbling blocks to the commercialization of a new technology and only reacts to information suggesting a risk from the new technology, product or service – often in a time frame too late to prevent harm to workers, consumers or to the environment.

Proactive standards development brings new challenges and opportunities. Proactive development of international interoperability standards (such as specification standards for material requirements) avoids elevating local standards developed by a single company to the regional level or national level. This would prevent expenditures on the subsequent costs of conversion to another standard and possible loss of economic leadership [23]. There is risk that with rapidly evolving technology, early lock-in on any overly specification-oriented standard can inhibit transition to superior performance standards in the future. Under these conditions, technological progress should be constantly monitored and standards adjusted to accommodate changes. Disproportionate influence by a single interest on the standards development process may lead to a suboptimal standard. Hence, balanced representation across all interest groups is critical.

The example of Genetically Modified Organisms (GMO) highlights the increasing influence of consumers over the market and promotes the shift from reactive to proactive risk management in the development of safety and health standards. GMO introduction into the food for human consumption initially occurred without identification of significant benefits to the consumer and without transparency about the safety of the new genetically-modified products to the consumer and the environment. The result of this lack of a proactive approach in looking down the risk road was public rejection of the technology which significantly hindered the development of an otherwise promising technology [53–56].

Under the conditions of proactive standards development, assuring the information quality of standards becomes critically important. A limited scientific basis for standards increases the role of expert opinions and makes standards development processes more vulnerable to influences of special interests. Tapping into the global pool of experts would make this process more robust and would ensure a more representative science-based consensus.

Proactive standards development also means that standards would be developed in parallel with standards validations. For example, a revision to an ASTM nanotechnology standard, ASTM E2490, Guide for Measurement of Particle Size Distribution of Nanomaterials in Suspension by Photon Correlation Spectroscopy, incorporates a large-scale inter-laboratory study that took place in 2008. The inter-laboratory study involved 26 laboratories conducting a total of 7,700 measurements of particle size distribution in three NIST Standard Reference Materials™ [57] and two solutions of dendrimers using several corroborative techniques including photon correlation spectroscopy. The results were factored into precision and bias tables that are now a part of the ASTM standard. As a reflection of the emerging nature of the field, the ASTM E2490 document is a practice guide rather than a prescriptive standard [58].

1.2.5 Emerging Development of Nanotechnology Standards

Standards development for nanotechnology reflects the new economic and political realities of the twenty-first century. The desire to guide the development of an emerging technology, and to proactively assess and manage any risks arising from that technology at the earliest opportunity highlights the challenging conditions under which nanotechnology standards are being developed. While the electronics industry has been at the forefront of proactive approach to standards development and IEEE coined the term "anticipatory" standards to describe standards produced well before the products they concern are commercialized, nanotechnology standards development has brought proactive standards development into the main stream and has become a testing ground for this approach.

Nanotechnology builds on achievements in a broad range of scientific and technological research since Richard Feynman first promoted the concept of working at the nanoscale [59]. Thus, a host of standards was rapidly developed for nanoscale objects, phenomena and techniques prior to the launch of nanotechnology-specific initiatives across the world. Some existing standards relevant to nanoscale measurement were established by "pre-nanotech" standards developing committees and include surface chemical analysis, sample preparation, microbeam analysis, material characterization and workplace air quality (for a more detailed list of existing and planned nanotechnology-related standards please refer to Annex C and D of Ref. [60]).

The launch of national nanotechnology programs in the first 5 years of the twenty-first century was followed by establishing nanotechnology technical committees and working groups in major standards developing organizations. Unlike the traditional structure of standards development around specific application areas, umbrella committees were formed to cover nanotechnology as a whole, which reflects the nascent nature of nanotechnology and the desire to guide its development.

A brief account of major milestones in national standards development starts in December of 2003 when China established a United Working Group for nanomaterials standardization and published the first Chinese industry nanotechnology standards in 2004. In May of 2004, the UK established NTI/1 national committee on nanotechnology. In the USA, ANSI at the request of the Office of Science and Technology Policy in the Executive Office of the President established in August, 2004 an ANSI Nanotechnology Standards Panel to coordinate nanotechnology standards development in the USA [61]. In November, 2004 Japan established a study group for nanotech standardization. And in November, 2005, the European regional standardization body, European Committee for Standardization (CEN), established CEN TC 352 Nanotechnologies.

Private standards developing organizations without national body membership on the international level were first to establish nanotechnology committees. In 2002, the IEEE Nanotechnology Council was formed as a multidisciplinary group to advance and coordinate the many nanotechnology scientific, literary and educational endeavors within the IEEE. The Council supports nanotechnology-related lectures, symposia and workshops, publishes the "IEEE Transactions on

Nanotechnology" and other periodicals, and sponsors nanotechnology standards [62, 63]. The IEEE Nanotechnology Council focuses on creating standards to aid commercialization, technology transfer and diffusion into the market including standards in nanoelectronics device design and characterization and quality and yield in high volume manufacturing.

The ASTM International Technical Committee E 56 on Nanotechnology was formed in 2005 [64]. Its work is organized into four technical subcommittees: "Informatics and Terminology," "Characterization: Physical, Chemical, and Toxicological Properties," "Environment, Health, and Safety," and "International Law and Intellectual Property."

Private international standards developing organizations with national body membership soon followed. ISO established a technical committee for nanotechnologies, TC 229, in June, 2005. The technical committee is structured around four working groups on "Terminology and Nomenclature," "Measurements and Characterization", "Health, Safety and the Environment", and "Material Specification". This technical committee also established several task groups aimed at exploring nanotechnology standards development and consumer and societal dimensions and sustainability. As of the eighth meeting held in June, 2009, ISO TC 229 brought together 32 participating member countries and eight observer countries, and the membership keeps growing.

In 2006, IEC established TC 113 in the field of nanotechnologies. This technical committee has three working groups: two joint with ISO TC 229 on "Terminology and Nomenclature" and "Measurements and Characterization" and the third on "Performance Assessment." As of August 14, 2009 the committee has 15 participating countries and 15 observers.

OECD was one of the first major international treaty organizations to establish nanotechnology groups. In 2006, OECD's Council established the Working Party on Manufactured Nanomaterials (WPMN) as a subsidiary body of OECD's Chemicals Committee [65]. The WPMN, in its turn, established nine steering groups to undertake specific tasks including development of guidance on toxicity testing and on exposure measurements and mitigation. In 2007, the OECD Committee on Science and Technology Policy established a Working Party on Nanotechnology (WPN) to look at economic and policy issues. WPN organized its activities into six project areas including policy dialogue, statistical framework for nanotechnology, and monitoring and benchmarking nanotechnology developments [66]. OECD has been especially active in the area of exposure assessment and mitigation for the nanotechnology workplace [65].

Although a number of agencies within the UN family of agencies have indicated their interest in nanotechnology, only a few exploratory and information-exchange activities have been initiated. Examples of early UN activity include: (1) a joint WHO/FAO expert meeting exploring safety of implications of applications of nanotechnologies in food and agriculture held in June 2009 [67]; (2) a UNESCO conference exploring ethical and social aspects of nanotechnology held in June 2007 [59]; and (3) series of workshops on risks of nanomaterials organized by United Nations Institute for Training and Research (UNITAR) [68]. Since 2006, the WHO Global Network of Collaborating Centers in Occupational Health has included nanotechnology projects in the WHO Network Workplan aimed at advancing the Network's Global Plan of Action [69].

Since nanotechnology covers a very broad range of applications, and an increasing number of international standards developing organizations are initiating activities in this field, there is a need for close coordination both within and between standards developing organizations. For example, as of June 2009 ISO TC 229 established 25 internal liaisons including liaison with REMCO, IEC TC 113 and CEN TC 352. There are also six external liaisons with outside organizations (OECD, EC Joint Research Center, Versailles Project on Advanced Materials and Standards, Asia Nano Forum, Bureau International des Poids et Mesures, European Environmental Citizens Organization for Standardization). Similarly, OECD WPMN recognized the importance of coordination with other standards developing organizations and outlined coordination activities in its roadmap for 2009 and 2010 [70].

In addition to bilateral agreements, a multi-stakeholder forum was convened by the USA government's National Institute of Standards and Technology, in February 2008 to further promote a dialogue among the standards developing organizations active in nanotechnologies standardization to identify standards needs related to nanotechnology [60]. At the 2008 meeting, participants agreed to develop: (1) a discussion forum to align information and developments from the different standards developing organizations; (2) a centrally maintained, searchable and freely accessible repository of information on existing standards and standardization projects in the field; (3) a database of existing measurement tools and new tools needed; (4) a searchable database covering definitions and terminology from all sources [60].

The structure of nanotechnology standards development committees follows broad application areas and was adopted in this book. Thus, eight chapters provide state-of-the-art review articles on progress in major standards developing areas: *Nomenclature & Terminology, Reference materials, Metrology, Performance standards, Application measurements, Implication measurements, Biological activity testing, and Health and safety*. Each chapter summarizes active areas of national and international standards development, together with its supporting knowledge base and emerging issues. The book also puts standards development in the context of legal international requirements and application of international standards to national governance structures in a dedicated chapter on *Legal considerations*. Specifically, this chapter discusses how nanotechnology standardization provides a common platform for addressing environmental, occupational and consumer implication issues and enables trade across differing national regulatory frameworks.

1.3 Conclusion

Throughout human history, standards have been crafted to enhance man's relationship to the laws of nature, to facilitate commerce, to promote technological innovation, to ensure the safety and health of workers, consumers and the environment, and to advance the standard of living for all mankind.

As the means of communication have improved, the range of stakeholder experts who develop standards, as well as the national and international reach of standards, has grown from localities, to regions, to nations and to the world, and from small trade groups to the global economy. The informational foundation for standards development has changed, which permits standards development to mature from a reactive mode, where well-established knowledge is used to set a standard, to a proactive mode, where the knowledge is generated in parallel with standards development, where the standards development guides and promotes the advance of technological innovations, and where precautionary approaches are put into place when risk information has yet to be definitely generated. Lastly, a global risk governance process is emerging where the pace of national governmental mandatory standards is being eclipsed by international, private sector, voluntary standards development.

References

1. ANSI. USA Standards System – Today and Tomorrow. e-Learning course. http://www.standardslearn.org (2009). Accessed 22 January 2010
2. NIST. A Walk through Time. http://physics.nist.gov/GenInt/Time/time.html (1995). Accessed 7 July 2010
3. Davies, G.: A History of Money from Ancient Times to the Present Day. University of Wales Press, Cardiff (1996)
4. Cardarelli, F.: Encyclopaedia of Scientific Units, Weights and Measures. Their SI Equivalences and Origins. Springer, London (2005)
5. Martin, H.-J.: The History and Power of Writing. University of Chicago Press, Chicago, IL (1995)
6. Hornung, E.: The Secret Lore of Egypt. Its Impact on the West. Cornell University Press, Ithaca, NY (2002)
7. Roberts, C.A.: The Food Safety Information Handbook. Oryx Press, Wesport, CT (2001)
8. German Beer Institute. German Beer Primer for Beginners. http://www.germanbeerinstitute.com/beginners.html (2008)
9. Weyrauch, T.: Craftsmen and their Associations in Asia, Africa and Europe. VVB Laufersweiler, Wettenberg (1999)
10. Smith, A.: An Inquiry into the Nature and Causes of the Wealth of Nations, 5th edn. Methuen & Co., Ltd, London (1904)
11. ANSI. Through history with standards. http://www.ansi.org/consumer_affairs/history_standards.aspx?menuid=5
12. ASTM International. ASTM: 1898–1998; a century of progress. http://www.astm.org/IMAGES03/Century_of_Progress.pdf
13. http://www.astm.org/ABOUT/aboutASTM.html
14. http://www.ieee.org/web/aboutus/history/index.html
15. McWilliam, R.C.: BSI: The First Hundred Years. Thanet Press, London (2001)
16. http://www.gost.ru
17. http://www.whitehouse.gov/omb/rewrite/circulars/a119/a119.html
18. http://osha.gov/pls/oshaweb/owadisp.show_document?p_table=MOU&p_id=323
19. http://www.sac.gov.cn/templet/default/
20. The United Kingdom Government and BSI. Memorandum of understanding between the United Kingdom Government and the British Standards Institution in respect of its activities as the United Kingdom's National Standards Body. http://www.berr.gov.uk/files/file11950.pdf (2002). Accessed 3 February 2010

21. Trebilcock, M.J., Howse, R.: The Regulation of International Trade, 3rd edn. Routledge, New York, NY (2005)
22. Kish, G.: A Source Book in Geography. Harvard University Press, Harvard, MA (1978)
23. Abbott, K.W., Snidal, D.: International 'standards' and international governance. J. Eur. Public Policy **8**(3), 345–370 (2001)
24. http://www.un.org/aboutun/unhistory/
25. http://www.un.org/en/documents/charter/
26. http://www.codexalimentarius.net/web/index_en.jsp
27. WHO/FAO: CODEX Alimentarius Commission. Procedural Manual, 18th edn. WHO/FAO, Rome (2008)
28. International Health Conference. Constitution of the World Health Organization. http://apps.who.int/gb/bd/PDF/bd47/EN/constitution-en.pdf (1946). Accessed 3 February 2010
29. http://www.who.int/rpc/expert_panels/EAP_Factsheet.pdf
30. WHO. Global Programme on Evidence for Health Policy. Guidelines for WHO Guidelines. World Health Organization, Geneva (2003) (EIP/GPE/EQC/2003.1). http://whqlibdoc.who.int/HQ/2003/EIP_GPE_EQC_2003_1.pdf. Accessed 3 February 2010
31. Global Health Watch 2: An Alternative World Health Report. Zed Books, Ltd., New York, NY (2008). http://www.ghwatch.org/ghw2/ghw2pdf/ghw2.pdf
32. Oxman, A.D., Lavis, J.N., Fretheim, A.: Use of evidence in WHO recommendations. Lancet **369**, 1883–1889 (2007)
33. http://www.oecd.org
34. http://www.oecd.org/department/0,3355,en_2649_34381_1_1_1_1_1,00.html
35. OECD: Convention on the Organization for Economic Co-operation and Development. OECD, Paris, France (1960)
36. Ozmańczyk, E.J.: Encyclopedia of the United Nations and International Agreements, vol. 2, 3rd edn. Routledge, New York, NY (2004)
37. Kuert, W.: Founding of ISO. http://www.iso.org/iso/founding.pdf
38. http://www.astm.org/
39. http://standards.ieee.org/
40. WTO. The WTO agreement on technical barriers to trade. http://www.wto.org/english/tratop_e/tbt_e/tbtagr_e.htm. Accessed 8 November 2009
41. OECD: Regulatory Reform and International Standardization. OECD, Paris (1999). TD/TC/WP(98)36/FINAL
42. Murashov, V., Howard, J.: The US must help set international standards for nanotechnology. Nat. Nanotechnol. **3**, 635–636 (2008)
43. Dern, D.P.: The Internet Guide for New Users. McGraw-Hill, Inc., New York, NY (1994)
44. Hayes, B.: Cloud computing. Commun ACM **51**(7), 9–11 (2008)
45. ISO. ISO concept database – user guide. Release 1.0. ISO, Geneva (2009). http://www.cdb.iso.org. Accessed 8 November 2009
46. Fletcher, D.: Wikipedia. Time 2009, August 18. http://www.time.com/time/business/article/0,8599,1917002,00.html
47. http://en.wikipedia.org/wiki/Wikipedia:About
48. Giles, J.: Internet encyclopaedias go head to head. Nature **438**, 900–901 (2005)
49. http://tc229wiki.ansi.org/tiki-view_articles.php
50. http://goodnanoguide.org
51. Van Noorden, R.: The science of Google Wave. Nature (2009), Published online 24 August 2009. doi:10.1038/news.2009.857
52. ISO: Consultation for the ISO Strategic Plan 2011–2015. ISO Central Secretariat, Geneva (2009)
53. Bradford, K.J., Van Deynze, A., Gutterson, N., Parrott, W., Strauss, S.H.: Regulating transgenic crops sensibly: lessons from plant breeding, biotechnology and genomics. Nat. Biotechnol. **23**, 439–444 (2005)
54. International Service for the Acquisition of Agri-Biotech Applications. Brief 37-2007: Executive Summary – Global Status of Commercialized Biotech/GM Crops. ISAAA (2007). www.isaaa.org/resources/publications/briefs/37/executivesummary/

55. Fox, J.L.: Puzzling industry response to ProdiGene fiasco. Nat. Biotechnol. **21**, 3–4 (2003)
56. Paarlberg, R.: Starved for Science: How Biotechnology Is Being Kept Out of Africa. Harvard University Press, Harvard, MI (2008)
57. NIST. NIST reference materials are 'gold standard' for bio-nanotech research. NIST Tech Beat 8 Jan. 2008
58. http://astmnewsroom.org/default.aspx?pageid=1840
59. www.its.caltech.edu/~feynman/plenty.html
60. ISO/IEC/NIST/OECD. ISO, IEC, NIST and OECD International workshop on documentary standards for measurement and characterization for nanotechnologies, NIST, Gaithersburg, MD, 26–28 February 2008. Final report, June 2008. http://www.standardsinfo.net/info/livelink/fetch/2000/148478/7746082/assets/final_report.pdf
61. http://www.ansi.org/news_publications/news_story.aspx?menuid=7&articleid=735
62. Rashba, E.: Nanotechnology standards initiatives at the IEEE. J. Nanopart. Res. **6**(1), 131–132 (2004)
63. http://ewh.ieee.org/tc/nanotech/
64. http://www.astm.org/COMMIT/COMMITTEE/E56.htm
65. Murashov, V., Engel, S., Savolainen, K., Fullam, B., Lee, M., Kearns, P.: Occupational safety and health in nanotechnology and Organisation for Economic Cooperation and Development. J. Nanopart. Res. **11**, 1587–1591 (2009)
66. OECD. OECD Working Party on Nanotechnology (WPN): Vision Statement. OECD, Paris, France (2007). http://www.oecd.org/sti/nano
67. http://www.fao.org/ag/agn/agns/expert_consultations/Nanotech_EC_Scope_and_Objectives.pdf
68. Strategic Approach to International Chemicals Management. Report of the International Conference on Chemicals Management on the Work of Its Second Session. SAICM/ICCM.2/15, SAICM, Geneva, 2009. http://www.saicm.org/documents/iccm/ICCM2/ICCM2%20Report/ICCM2%2015%20FINAL%20REPORT%20E.pdf
69. WHO. Workplan of the Global Network of WHO Collaborating Centers for Occupational Health for the period 2009–2012. WHO, Geneva (2009). http://www.who.int/occupational_health/network/priorities.pdf
70. OECD. Manufactured Nanomaterials: Roadmap for Activities During 2009 and 2010. ENV/JM/MONO(2009)34. OECD, Paris, France (2009). http://www.olis.oecd.org/olis/2009doc.nsf/ENGDATCORPLOOK/NT00004E1A/$FILE/JT03269258.PDF

Chapter 2
Current Perspectives in Nanotechnology Terminology and Nomenclature

Fred Klaessig, Martha Marrapese, and Shuji Abe

2.1 Introduction

At the time of writing this chapter, early in 2010, several reports have been issued that differ in definitions used for nanotechnology, which is not unusual considering the large number of conferences, reports, papers and presentations given each year on this subject. It is in fact very difficult to follow developments in this field, and the multidisciplinary nature of nanotechnology almost invites a similar multiplicity of definitions as each specialty (or scientific discipline) adjusts to the new findings of what is a dynamic research effort. However, the same dynamism leads to ambiguity in meanings and to uncertainty in the overall impact this field will have when products are commercialized. In this chapter, we will be visiting the several dimensions, societal, governmental and technical, and thereby highlighting the challenges facing terminology and nomenclature efforts.

One example of the public dialog, and one very timely to this article, is the recent publication by the U.K. House of Lords Science and Technology Committee titled, "**Nanotechnologies and Food**" [1]. The 12 panel members have distinguished public careers, in many cases as Members of Parliament, and came to the recommendation:

> …We recommend … that any regulatory definition of nanomaterials … not include a size limit of 100 nm but instead refer to 'the nanoscale' to ensure that all materials with a dimension under 1000 nm are considered.

The recommendation is that the term *nanoscale* have an upper boundary of 1,000 nm for the purpose of food regulations, rather than the ISO and ASTM International determinations that scientific usage is 100 nm. The European Union in recent legislation regarding cosmetics labeling remains with the 100 nm upper boundary, but also includes materials of unspecified size that contain nanoscale

F. Klaessig (✉)
Pennsylvania Bio Nano Systems, LLC, Doylestown, PA, USA
e-mail: fred.klaessig@verizon.net

V. Murashov and J. Howard (eds.), *Nanotechnology Standards*,
Nanostructure Science and Technology, DOI 10.1007/978-1-4419-7853-0_2,

components. And, SCENIHR, an advisory body to the European Commission on public health for new and emerging risks, is evaluating a surrogate metric for identifying nanoscale materials based on a specific surface area of greater than 60 m^2/g, which is considered a property characteristic of the nanoscale. What is apparent from these separate approaches is that the term *nanoscale* remains in flux and has multiple meanings in the context of public policy.

While governmental groups responding to public sentiment for a regulatory policy are migrating towards sizes larger than 100 nm for nanoscale, the materials sciences community has tended to lower sizes when referring to the unique, novel and unexpected properties to be associated with nanoscale materials. One well respected group recently suggested 30 nm as the size below which unique, size-dependent properties are to be observed, especially those associated with quantum confinement. An initial tendency of those concerned primarily with biological sciences to favor a larger concept for nanoscale, up to 1,000 nm, while those concerned primarily with materials science pursue smaller sizes, was also noted in a recent workshop on Nanomedicine and Terminology sponsored by the American National Standards Institute (ANSI). Colleagues in the medical fields were familiar with dimensional scales from biology, e.g. size exclusion phenomena involving cell and organ processes, which are significant to biological responses and extrinsic to the nanoscale material. The materials scientists were more focused on the intrinsic, system-independent characteristics. What is apparent from this scientific dialog between extrinsic and intrinsic properties is that the scientific community has not yet come to agreement on the characteristics of nanoscale materials that are to be emphasized for the purposes of definitions.

Ambiguity in terms of competing definitions is not surprising in a rapidly developing technology. Scientists in a new field, especially one as vibrant and having as many practitioners as nanotechnology, both generate new terms and borrow others freely from more established disciplines. There are, nevertheless, nuances that are not always explicitly resolved until more data are generated. In the book "**Structures of Scientific Revolutions**," T.S. Kuhn [2] uses the concept of incommensurability to describe two groups, one using the older paradigm and one the newer, having difficulties in communicating when using seemingly common expressions. The two paradigms differ in tools, terms and descriptive models, and communications suffer. Nanotechnology, in particular, experiences the difficulty that the properties of nanoscale materials are intermediate to those associated with molecules and also to those associated with the bulk, extended phase. There is no mathematical formulation to follow, and at best, there are referrals to quantum confinement or to the high surface-to-volume ratios for explanations. The intermediate position between molecules and bulk corresponds to philosophical controversies surrounding emerging properties, reductionism and mereology. Thus nanotechnology, as a new field operating in a size range that places collective properties into question, itself faces unique descriptive challenges.

An article such as this one faces several limitations. As we write, each of the organizations involved continues in their efforts to set standards, ballot new terms, initiate new work items or, for some, re-visit earlier documents to maintain currency.

Similarly, it is difficult to keep pace with, no less anticipate, the magnitude of new commercial products that will eventually emerge from nanotechnology. Out of pragmatism, and realizing that many readers may be new to this subject area, we survey the field using three simple questions that allow us to differentiate among existing efforts, as well as give guidance to future directions. They are:

1. What is the nanoscale?
2. What properties are associated with nanoscale materials?
3. What is a nanomaterial and are there distinguishable categories of nanomaterials?

One definition of judgement, this one from Webster's Dictionary [3], is "*the power of arriving at a wise decision or conclusion on the basis of indications and probabilities when the facts are not clearly ascertained*." Each of the groups mentioned in this article is attempting to apply judgement to a rather dynamic field in order to extract those elements most applicable to their immediate responsibilities (schema, paradigm, mission statement or statutory authority). Though the efforts of a broad collection of organizations are reviewed in this chapter for their reflection of societal concerns about the impact of nanotechnology, we, the authors, will primarily emphasize the efforts of standards developing organizations (SDOs). In these latter cases, the consensus-oriented methodology crosses both national boundaries and scientific disciplines and is more likely to offer a coherent set of terms, definitions and nomenclature from the perspectives of those scientific fields that generate nanotechnology. The resulting standards are more likely to reduce overall ambiguity and thereby enable effective outwards communications with the broader, public policy community and society.

2.2 Terminology

We should begin by explaining what is meant by a terminology and a nomenclature, which can be illustrated by the everyday experience of conducting an internet search. We have all noticed how the selection of *key words*, a change in their number, their order or in making substitutions, affects the results such as the number, the arrangement and the immediate utility of the returned entries, colloquially called "hits." The key words are an uncontrolled listing of terms in that the person conducting the search and those designing the web sites acted independently in selecting descriptors. If one selected key words from a *terminology*, a listing of terms that reflect usage in a selected topic area, then the returned search ("hits") should be improved (fewer in number, more authoritative and more pertinent to the inquiry). A *terminology* is a list of terms used in a field, which means the person searching and those who operate web sites are more likely to be using the same set of key words.

If one adds a definition to each term in the terminology, one has a *vocabulary* or *glossary*. A well respected glossary, such as a dictionary, influences usage so that the terms are not only current, but carry a similar meaning to all practitioners following the vocabulary. In the field of library science, there is the concept of a *controlled vocabulary*. It is used to index (catalogue) information, it is structured, and it is adjusted over time as usage changes. The purpose of a controlled vocabulary is the ability to retrieve information (such as a book or a study) even though time has passed to the point that the subject field has changed. This allows two searches separated in time to find the same item. *Ontology*, a form of controlled vocabulary incorporating associative relationships among the terms, is also used in federated data bases to allow meaningful computer searches. A *nomenclature* is a system of terms that is combined with pre-established rules in order to name or classify an item in a consistent and unique manner. A nomenclature system aids in proper cataloguing and retrieval of information in a manner similar to a controlled vocabulary.

There are several recognized international standards developing organizations that develop terminologies, vocabularies and nomenclature systems, as well as test techniques, material specifications and business processes of interest to commerce. In this chapter, we will be emphasizing two of them: ASTM International and ISO. For both, there exist committees dedicated to a broad range of nanotechnology issues, and they are governed by rules regarding consensus, voting and representation. It is primarily this focus on nanotechnology that favors their work in terms of developing an internationally recognized and accepted set of terminologies and accompanying definitions.

There are other organizations (we will refer to them as non-SDOs) that have offered definitions for individual nanotechnology terms within the context of their primary interests like a statutory responsibility, a mission statement or a reflection of other business and societal interests. We recognize that these organizations have their own informed internal processes for their proposals and we make our distinction primarily on the basis that the SDOs are more likely to establish a coherent, structured approach to nanotechnology terminology and nomenclature.

2.2.1 Non-SDO Sources

Our first question is, "What is the nanoscale?" Table 2.1 lists the suggested upper boundary for the term *nanoscale* along with the organization (and reference) and noteworthy qualifications. It is arranged in terms of increasing upper limit.

There is an order of magnitude across the size range plus very specific justifications for the intermediate values of 200, 300 and 500 nm, clearly indicating that there are many informed voices participating in this public dialog. There are of course provisos to take into consideration. The Chatham House and Swiss Re reports for example do not focus on the scientific underpinnings of nanotechnology, but rather on the regulatory framework and general risk ramifications the topic poses, and in doing so, the authors make some general references to sizes of

Table 2.1 Compilation of recommended upper limits suggested by different organizations

Upper limit (nm)	Source [ref. no.]	Comment(s)
100	ISO [4]	"Approximately" qualifies size range; lower limit is 1 nm (see explanatory note 2)
100	ASTM Int. [5]	"From approximately" qualifies size range; lower limit is 1 nm
100	Royal Society [6]	Lower limit is 0.2 nm (size of atoms)
100	SCENIHR [7]	"Of the order of" qualifies 100 nm; recent suggestion to calculate size from BET surface area
100	Sci. Cmt. on Consumer Products [8]	"Of the order of" qualifies 100 nm
100	ETC Group [9]	"Below about" qualifies 100 nm
100	Swiss Re [10]	"Smaller than" qualifies 1–100 nm; alternative sizes mentioned: <200 nm evading phagocytosis; <300 nm associated with particle migration and Peyer's plaques
200	Soil Association [11]	Intended as the mean of a particle size distribution with the smallest particle being <125 nm
200	Defra [12]	Basis of data call in with qualification to two or more dimensions (no 1-d flakes or coatings included)
300	Chatham House [13]	A suggested limit for regulatory purposes
300	Friends of Earth [14]	Between 0.3 and 300 nm; might be larger if size is important to function or to toxicity
500	Swiss Federal Office of Public Health [15]	Categorizes particles into size ranges with boundaries at 100 and 500 nm
1,000	House of Lords Science Committee [1]	Ingested particles appear to be the basis for selecting 1,000 nm; see page 111 of vol. II

biological importance. The UK's Soil Association, which has a firmly established process for creating standards, aligned their definition with a pragmatic interpretation of particle size distribution measurements that the other sources probably assumed was a separate consideration. Though the SDOs and the several scientifically oriented societies and committees tend to quote a 100 nm upper limit, we must also note differences among them in terms of lower limits and in the use of adverbial qualifiers such as 'approximately' or 'of the order of'.

Of particular note is the recent publication by the U.K. House of Lords Science and Technology Committee. The 12 panel members do not purport to be a scientific body, but with distinguished public careers (as Members of Parliament, government civil servants and individuals with a science background), they are able to evaluate public sentiment when addressing public policy. In the case of this panel's report, the reader also has access to the individual testimony of those representing a broad array of scientific bodies, industry trade groups, interested associations as well as governmental agencies. We can assume that the report essentially extracts what in the panel's judgement was a useful basis for public policy regarding food. The Panel noted that witnesses expressed definitions of varying clarity for terms such as nanotechnologies, nanomaterials, nanoparticles ([1], Q474) and nanoscale properties ([1], Q487), leading the panel to its recommendation for a 1,000 nm upper boundary. (There is a separate discussion of definitions in regulatory decisions to be found in Chapter 10 of this book, "Nanotechnology Standards and International Legal Considerations").

The organizations suggesting limits above 100 nm expressed a more biologically oriented reasoning for doing so. The sizes of viruses, quoted as 10–300 nm, or specific mechanisms of cell entry, such as endocytosis, are mentioned. And, the two advisory committees to the European Commission, though remaining with the 100 nm limit, also mention biological mechanisms as a basis for taking particular caution. In the case of the House of Lords Committee report, particular note was taken of Professor Jonathan Powell's work, who was in fact the only witness to mention 1,000 nm, in lieu of the size suggestions from other participants.

The physical and materials sciences literature focuses on the intrinsic properties exhibited by a material with examples being density, melting point, refractive index, and other properties that are relatively independent of the immediate environment. Dictionary definitions of intrinsic properties tend to emphasize essential characteristics, but here we would rather emphasize the concept that these are properties without reference to the surrounding environment. The SDOs, where primarily materials scientists participate, conform to the general scientific literature. Colleagues concerned about the important environmental, health and safety aspects tend to emphasize the extrinsic properties that a nanoscale material may exhibit in biological settings, which the House of Lords Committee report expresses very well, "*A change in functionality, meaning how a substance interacts with the body, should be the factor that distinguishes a nanomaterial from its larger form within the nanoscale.*"

The distinction between intrinsic and extrinsic properties recently arose at a Nanomedicine and Terminology workshop co-sponsored by the American National Standards Institute and the Chemical Heritage Foundation. Participants had academic, governmental and industrial backgrounds, but the more obvious distinction was between those concerned with quantum confinement (materials scientists) and those concerned with elimination by the kidney or passing the blood brain barrier (biological scientists). When addressing nomenclature, the physical scientists tended to work from the particle center outwards, while the biological scientists tended to do the reverse. It is likely that both groupings are correct in their respective areas.

Our second question is, "What properties are associated with nanoscale materials?"

In a recent peer reviewed article [16], using a similar intrinsic and extrinsic categorization as mentioned above, the physical scientist authors suggested that the onset for size-dependent properties was more likely to be observed at 30 nm, lower than the conventionally quoted 100 nm. As a contrast, Professor Powell's literature studies suggest sizes up to 2,000 nm interact with M-cells and Peyer's patches [17]. It should be noted that there has been a greater emphasis on respiratory studies and the lung in the toxicological literature than for ingestion and the GI tract [18], which may explain why the House of Lords Committee viewed Professor Powell's comments as most significant to their remit of food policy. It should be noted, too, that many other fields of ecotoxicology have not yet established a broad body of knowledge on nanoscale materials and may in the future propose other boundaries, both larger and smaller, for the nanoscale. Caution until more is known can lead some workers to a working preference of 1,000 nm [19, 20] or, as in the caNanoLab glossary [21], to definitions without a specified size range, "*A nanoparticle is defined as a small, stable particle, whose size is measured in nanometers*." Here we wish to emphasize that the properties various organizations or scientific disciplines associate with nanoscale materials can be quite varied.

In general, the several non-SDO groups have been primarily concerned with the uncertainty surrounding the risk analysis or risk assessment of nanoscale materials rather than specific properties. Their comments are frequently drawn to the absence of data. Hence, their frequent use of analogies to viruses or to physiologic particles when discussing the potential for nanoscale materials to exhibit unsuspected properties. With the non-SDO groups, the emphasis is how the nanoscale material fits with a substantive, pre-existing commitment, whether to a statutory responsibility, a mission statement or to a scientific field. When these groups are considered in their totality, they reflect the societal implications of commercial applications utilizing nanoscale materials.

In Europe, there has been legislation (the cosmetics directive [22]) and there is proposed legislation for novel foods [23] that attempt to clarify the significant attributes, which for cosmetics include:

1. Insoluble or biopersistant [22]; and
2. Those related to the large specific surface area of the materials considered [23]; and/or
3. Specific physico-chemical properties that are different from those of the non-nanoform of the same material [23].

The issues of insolubility and biopersistence for cosmetics presumably address the nanoscale material's residence times in the commercial preparation (shelf life) and later, after intended product use, under environmental and physiological conditions. The remaining attributes re-phrase the more conventional concepts of size (surface area) and the expectation properties of a material's nanoscale form might differ in an unexpected fashion from those of the larger form. Rephrasing size through surface area or surface-to-volume ratios effectively serves to raise the size limit to the term *nanoscale* or to the spectrum of products viewed as nanomaterials.

Our third question is, "What is a nanomaterial and are there distinguishable categories of nanomaterials?"

In utilizing size considerations of a biological nature when defining *nanoscale*, the House of Lords Committee's suggestion effectively defines a nanomaterial as something less than 1,000 nm in size that reaches a biological system and elicits a different response than the non-nano-form of the same material. There are many references in the report regarding the difficulty in defining *nanomaterial*, which the panel eventually places into recommendation 11 ([1], page 76) that legislation include "*workable definitions of nanomaterials and related concepts.*" We should recognize that by using a 1,000 nm upper boundary for the nanoscale, the panel accepts that it is casting a wide net for potential nanomaterial candidates, including an element of caution and safety. A similar expression of difficulty in defining a nanomaterial is found in a recent EPA expert report on nano-silver, where the panel comments, "*A critical issue that must be clarified is use of the terminology "nano". The common definition is one that often includes <100 nm in one dimension and poses a unique property. For standardization, the unique property of nanosilver should be established, as well as for aggregates of nanosilver or nanosilver incorporated via binders.*" ([24] page 38). Similar comments would probably apply to the other organizations in Table 2.1 that have set upper boundaries at 200, 300 and 500 nm when connecting their definitions of nanoscale to nanomaterial or vice versa.

In stressing size as the primary defining element to a nanomaterial, there remain some open questions on what is being measured at the nanoscale. The House of Lords Committee report includes solid particles, emulsion micelles and apparently biomolecules, e.g. enzymes, as "materials" whose size is germane. The last point of biomolecules requires some elaboration. Several of the witnesses certainly mentioned "*engineering internal structures at the molecular level (i.e. nanoscale)*" ([1], vol. II, page 321) or "*manipulation at the nanoscale*" ([1], vol. II, page 133), eventually leading to descriptions of ice cream or mayonnaise as potential nanomaterials. In an other forum, a scientific committee responding to EFSA-Q-2007-124a stated, "*Food and feed may contain components that have internal structures that individually could be present at the nanoscale, e.g. naturally occurring molecules, micelles or crystals.*" ([25], page 8). The concept of biomolecules as nanoscale entities is controversial, as molecules are generally excluded from consideration when using the scope statement of TC 229 ("*Utilizing the properties of nanoscale materials that differ from the properties of individual atoms, molecules, and bulk matter*" [26]). But, again, this is a point of diverse opinions and, like extrinsic and intrinsic properties, may reflect the different worldviews of physical and biological scientists.

The purpose here, though, of pointing out this unclear situation with biomolecules is that including them as forms of nanomaterial tends to bring a large component of the traditional human diet into the discussion of nanotechnology. This step, in turn, has a cascading effect, leading to differentiations being made among natural, manufactured, engineered and incidental categories of what might otherwise be the same chemical substance, e.g. silica. Hence, the House of Lords Committee report excludes natural nanomaterials except if they have been selected or processed intentionally as on page 51 ([1], vol. I):

> We recommend that, for regulatory purposes, any definition of 'nanomaterials' should exclude those created from natural food substances, except for nanomaterials that have

been deliberately chosen or engineered to take advantage of their nanoscale properties. The fact that they have been chosen for their novel properties indicates that they may pose novel risks.

Similar conclusions can be found in [7, 23, 25]. Having participated in the standardization process from primarily the technical viewpoint, albeit from a materials science one, recommendations such as these go beyond intrinsic properties and make selecting a nanoscale component for commercial use the crux is defining novel properties, when in the case of mayonnaise, the intention may have been the objective of lowering fat content. Clearly, the issue of molecules is a point of controversy for all to consider.

Though the two SDOs will be emphasized for their focus on nanotechnology within a broad field of concepts, it should be noted that the SCENIHR references are quite extensive themselves in terms of offering self-consistent definitions and of cautioning others about creating new terms using the prefix *nano*. The terms given definitions are (see the first citation in [7]) (Table 2.2):

Table 2.2 SCENIHR terms categorize by nano-prefix and properties

Terms	Properties
• Nanoscale	• Agglomerate
• Nanostructure	• Aggregate
• Nanomaterial	• Coalescence
• Nanocrystalline material	• Degradation
• Nanocomposite	• Solubilisation
• Engineered nanomaterial	
• Nanosheet	
• Nanorod	
• Nanotube	
• Nanoparticle	
• Nanoparticulate matter	

There is a general caution given to limit the proliferation of new terms using the prefix *nano* (see especially [7], Sect.3), which will be discussed more fully when reviewing ASTM International activities. Overall, the SCENIHR panel wrestled with the same issues outlined here, but emphasized a risk assessment relevance to their proposals. Nanoscale is not defined as a linear distance, but as a "*feature characterized by dimensions of the order of 100 nm or less*" ([7], Sect.3.3.3.1). Where definitions by other organizations are expressed in terms of a linear distance, the SCENIHR definition is a physical object or entity, and the standard dictionary connotations of the word *scale* are not present. Yet, the other sources have used adverbial qualifiers, such as approximate, and frequently connect the stated size range to material properties, such as note 1 to *nanoscale* in [4], "*For such properties the size limits are considered approximate*." The nanoscale is not like the Celsius scale or the Richter scale that can be calibrated independently. There is no standard nanoscale reference material for properties. In this light, the SCENIHR definition for nanoscale, though taking an unexpected form, is simply more explicit in having a material entity present at the outset, while still wrestling with clear upper and lower boundaries for *nanoscale*.

The SCENIHR panel recognized that "*Most of the concepts and behaviour patterns seen at the very small dimensions associated with nanotechnology are not new...*" ([7], page 3). When addressing properties, their focus is on those describing the fate of a nanoparticle, e.g. degradation, and not on those that might define a nanoparticle before its commercial use. Though there is a formal definition given for nanomaterial, the central concept is well expressed in 3.3.3.2 as, "*It is proposed that, as a general rule, if a material has distinctly different properties from the bulk material as a consequence of its occurrence as discrete entities (nanoparticles, nanosheets, nanorods or nanotubes) with one or more dimensions of 100 nm or less, it should be considered as a nanomaterial.*" The challenge, as for the SDOs to be reviewed below, remains one of categorizing nanomaterials by properties. Finally, there is some ambiguity regarding the status of molecules, which are discussed with the term *nanostructure*. The emphasis is placed on "discrete functional parts," but the document does not provide enough examples to know if a biomolecule, such as a protein molecule, has discrete parts due to different functional groups being present. In polymer chemistry, the question would be if block co-polymers are viewed as nanostructured, while homopolymers are not.

In a more recent SCENIHR publication [7], there is mention that 60 m^2/g surface area should be a defining criterion for a nanomaterial, in part to be certain that agglomerates and aggregates, even those much larger than 100 nm, are included. This specific surface area value aligns with that of a 100 nm solid sphere of unit density (1 g/cm^3). Any other filled shape of the same mass and density would have a larger specific surface area value. A correction for density is recommended between materials. The suggestion for a surface area criterion is mirrored in several NGO and trade group discussions. This specific surface area criterion is a workable concept for powders (particles, aggregates and agglomerates), but may require additional commentary to be applicable to solid particulates and micelles in dispersions (where BET measurements are not possible) or to molecular entities. Even with solids there will be issues with porous materials and whether a void or interconnected pores can exhibit unusual properties relative to the fluid's bulk extended phase.

2.2.2 ASTM International

Although we have referred frequently to ASTM International, it should be noted that the "Terminology for Nanotechnology" document E 2456-06 was developed in concert with several other SDOs:

- American Institute of Chemical Engineers
- American Society of Mechanical Engineers
- Institute of Electrical and Electronics Engineers

- Japanese National Institute of Advanced Industrial Science and Technology
- NSF International
- Semiconductor Equipment and Materials International

At the time of the E 2456-06 ballot, the E56 committee structure included (with the number of published standards provided in the parenthesis):

- Terminology & Nomenclature (1)
- Characterization (2)
- Environmental & Occupational Health & Safety (3)
- International Law & Intellectual Property (0)
- Liaison & International Cooperation (0)
- Standards of Care/Product Stewardship (1)

Since its founding in 2005, the committee has issued seven standards as categorized above. In addition, ASTM International has funded round-robin testing for one characterization method and for one toxicity test protocol, in line with its historical commitment to providing significance and use information for its standards. More information on the E56 committee can be accessed at http://www.astm.org.

In terms of the three questions we are posing, E 2456-06 addresses nanoscale and properties associated with nanoscale particles, but suggested definitions relating to *nanomaterial* did not complete the ballot process. Nanoscale encompassed "approximately" 1–100 nm. As with the Defra definition of *nanoscale*, the E56 definition of *nanoparticle* does not include 1-d nanoscale particles, which again reflects the dynamics of the balloting process combined with the complexity of differentiating a discrete 1-d nanoparticle from a nanoscale thick coating attached to a substrate.

One can generalize that the E56 terminology focuses on the properties to be associated with nanoscale particles and did not explore the full range of shapes that nanoscale particles may assume. In contrast, the first ISO standard is more detailed on the shapes, as will be discussed below, and has addressed property issues only more recently. Both SDOs found it necessary to create terms in order to avoid ambiguity. In the case of E56, this was the distinction between transitive and non-transitive nanoparticles. As already mentioned in this article, there is an expectation that nanoscale materials will exhibit novel or unique properties that cannot be extrapolated from measurements on the same material at the larger scale. The E56 Committee viewed those nanoparticles exhibiting a discontinuity in property (one that emerges solely at nanoscale size or does not extrapolate from larger sizes) as transitive, while those exhibiting no discontinuity are considered non-transitive. Though expressed in terms of properties, materials that have been in commerce for many decades are likely to be non-transitive as their development arose from progressively finer and finer milling steps as well as new synthesis technology. High specific surface area and optical scattering are given as examples of non-transitive properties. Finally, the early ballots for E56 contained the term *ultra-nanoparticle*, which was defined very close to the 30 nm onset value of [16]. This speaks to the fact that the participants in E56 were primarily physical scientists.

There is one fine point in the phraseology utilized in E 2456-06 that is similar to the earlier discussion on intrinsic and extrinsic properties. In this case, it is intensive and extensive properties. The definitions in E 2456-06 refer to intensive properties, which are those that are independent of the amount of material present. Surface area changes with the amount of material present, but specific surface area, m^2/g, does not if the sample is thoroughly mixed. Mass would be extensive, while density is intensive. The more recent use of 60 m^2/g as a defining criterion for nanoscale materials would be an example of an intensive property, but according to the E 2456-06 definition, it would be non-transitive, as there is no discontinuity when extrapolating from larger sizes.

Two other points in E 2456-06 are worthy of note. One refers back to the SCENIHR advice to be cautious with the number of terms with the prefix *nano*, and the second is a definition of the prefix *nano*. The Significance and Use section of the E 2456-06 lists criteria for introducing a new term, including: (a) currency in the scientific literature, (b) limiting changes to historial meanings to just those needed by nanotechnology practitioners, (c) giving precedence to established terms when there is overlapping usage among scientific disciplines and (d) delimiting to *in nanotechnology*. The major difference with the criteria in [4, 7] is relevance, where the SCENIHR committee members emphasize risk assessment or related purposes when reaching their decisions. A step towards avoiding ambiguity in E 2456-06 was taken by including terms from aerosol science to provide context (e.g., ultrafine particle).

The prefix *nano* is defined in three senses: (a) SI units; (b) small "*things*;" and (c) a set of concepts that must pertain to nanotechnology or nanoscience. The SCENIHR panel addressed a similar concern with the prefix *nano* in their discussions of the relative merits of the terms *nanostructured* (discrete functional parts) and *nanomaterial*. The SCENIHR panel's initial preference was to favor *nanostructured*, but they elected for reasons of scientific currency to remain with *nanomaterial*.

The E 2456-06 balloting process has been mentioned regarding *nanomaterial* and *ultra-nanoparticle*. The initial listing contained 78 entries, which became 13 in the final document. Responses to the initial ballot were very numerous and detailed, and it is noteworthy that some objections were eventually voted as "non-convincing." Two factors were influential during the balloting process. Firstly, the broad committee membership included many from the filler, pigment and material handling industries, who had worked with fine and ultrafine particle terminology and who expressed some reluctance to new terms replacing established ones. The second factor was that all of the terms were voted on as one single item; a step that drastically reduced the number of terms, but enforced consideration of related aerosol terms. The resulting terminology document has terms specific to nano-prefixed words that are delimited by "*in nanotechnology*" and that are provided context by concepts from aerosol and materials science.

ASTM International has added informatics to the title of its terminology subcommittee, and they are initiating activities for applying computational tools when connecting nanoparticle characterization to the results of biological testing. Returning to the introductory comments, a controlled vocabulary of terms and definitions is frequently used to retrieve information. A hierarchical controlled vocabulary containing terms, definitions and the relationships among the terms is referred to as an ontology. A familiar example would be a family tree. One example of a nanoparticle ontology is found at http://www.nano-ontology.org. A nanoparticle ontology can be used in generating an informatics capability, which would include creating, populating and maintaining a data base, in much the same way that a controlled vocabulary and index system is used in a library. Just as a person may be found in several family trees, so too can a topic area be an active research theme for several scientific disciplines. There is therefore great potential an informatics standard being a bridge among many independently maintained databases (a federated database), allowing for data mining, pattern recognition and machine learning. Furthermore, this can be done while retaining the independent relationships the topic has in the separate scientific fields, such as in the examples of intrinsic and extrinsic properties given in this article.

2.2.3 ISO TC 229

TC229 began in 2005 with 39 members (30 P- and nine O- members) and has since developed a structure of four working groups (WGs) with several task groups that support the Chair or, at times, individual Convenors. In early 2010, there are 19 liaisons with other ISO committees, one with IEC TC 113 and eight with external organizations. The Secretariat is British Standards Institute, and the four Working Groups are (Table 2.3):

Table 2.3 Working group structure of TC 229

Designation	Title	Convenorship
WG1	Terminology and nomenclature	Canada
WG2	Measurement and characterization	Japan
WG3	Health safety and environmental aspects of nanotechnologies	USA
WG 4	Material specifications	China

Though the primary commentary here will be with JWG1 (it is a Joint Working Group with IEC TC 113), there are issues within WG4 that will be mentioned: (a) specifications on the intrinsic properties of nanoscale materials and (b) coordination with CEN/TC 352 on the work item titled, “Manufactured Nanoparticles – Guidance on Labelling.”

One terminology standard has been issued [4], and three were undergoing administrative review for late 2010 publication. In view of the increasing importance of nanotechnology, ISO and IEC have agreed to a new, common numbering designation, the 80004-series, so that these standards can be readily recognized.

Table 2.4 Published and active work items for WG1

Designation	Title
ISO/TS 27687:2008 (to be re-issued as ISO/TS 80004-2)	Nanotechnologies – Terminology and definitions for nano-objects – Nanoparticle, nanofibre, and nanoplate
ISO/TR 12802	Nanotechnologies – Terminology – Initial framework model for core concepts
ISO/TS 80004-1	Nanotechnologies – Vocabulary – Part 1: core terms
ISO/TS 80004-2	Nanotechnologies – Vocabulary – Part 2: nano-objects – Nanoparticle, nanofibre, and nanoplate
ISO/TS 80004-3	Nanotechnologies – Vocabulary – Part 3: carbon nano-objects
ISO/TS 80004-4	Nanotechnologies – Vocabulary – Part 4: nanostructured materials
ISO/TS 80004-5	Nanotechnologies – Vocabulary – Part 5: bio/nano interface
ISO/TS 80004-6	Nanotechnologies – Vocabulary – Part 6: nanoscale measurement and instrumentation
ISO/TS 80004-7	Nanotechnologies – Vocabulary – Part 7: medical, health and personal care applications
ISO/TS 80004-8	Nanotechnologies – Vocabulary – Part 8: nanomanufacturing processes
ISO/TR 11360	Nanotechnologies – Methodology for the classification and categorization of nanomaterials

Table 2.4 is a listing of active terminology work items and the three issued standards exemplify the range of topics as well as the Committee's focus on the underlying scientific basis of nanotechnology.

Updated information on the Committee's structure, activities and standards can be obtained from the ISO website http://www.iso.org/iso/standards_development/technical_committees/list_of_iso_technical_committees/iso_technical_committee.

Before discussing TC 229 terminology activities, it should be noted that five of the ten initial work item project teams were influenced by publicly available specifications (PAS) used by BSI project leaders as starting points for deliberations. In those cases, the extent and depth of discussions were enhanced by this initial UK activity. The relevant documents are given in Table 2.5, though it must be pointed out that the eventual ISO document often differs greatly from the PAS (152 terms in PAS 71, compared to 12 terms in TS 27687:2008(E)). Yet, it may still be helpful for the reader to use the existing PAS for greater insight into the on-going deliberations on the remaining unpublished ISO work items.

PD 6699-1:2007, "Part 1: Good practice guide for specifying manufactured nanomaterials" is presently in WG4 as work item TS 12805. PAS 130:2007, "Guidance on the labelling of manufactured nanoparticles and products containing manufactured nanoparticles" is a CEN/TC 352 work item with the designation CEN ISO/DTS 13830.

Our first question is, "What is the nanoscale?" and, as in the E56 terminology standard, TC 229 defines the nanoscale as "approximately" 1–100 nm with an

Table 2.5 BSI and respective ISO document designations

BSI document	Related ISO document
PAS 71:2005	TS 27687:2008(E)
PAS 130:2007	CEN/TC 352
PAS 131:2007	ISO/TS 80004-7
PAS 132:2007	ISO/TS 80004-5
PAS 135:2007	ISO/TS 80004-8
PD 6699-1	ISO/DTS 12805

explanatory note regarding the lower end being advisable to avoid incorporation of single and small groups of atoms into the field.

Our second question is, "What properties are associated with nanoscale materials?" Here a distinction should be made, as the published standard does not address this question beyond a note to the term *nanoscale*. Here, the properties were not described, but rather the indication given that it is in this size range that one might expect to observe properties that are not simple extrapolations from larger size material. The "approximate" in the nanoscale definition is in consideration of the properties that do not extrapolate and in this sense can be viewed as emergent.

The TC 229 JWG1 discussions focused on establishing categories along with examples of the shapes nanoscale particles may take. These are intended to be foundational examples, though some do overlap in the details of the geometric descriptions. There is also the introduction of a new term, *nano-object*, to be an umbrella concept for all nanoscale objects. This step was taken, because the existing scientific literature has used nanoparticle to cover all shapes (rods, tetrapods, spheres), while the term *particle* is normally associated with a generally spherical shape. The decision taken was to use nano-object as the general term and to limit nanoparticle to 3-d shapes of a spherical nature. There is a hyphen in nano-object for reasons of English pronunciation. A hierarchy of terms and illustrations of the shapes are provided in the document's introduction. A simplified form of this hierarchy is:

- Nanomaterials may be nano-objects or nanostructured materials
- Nano-objects may be nanoparticles (3-d), nanofibres (2-d) or nanoplates (1-d)
- Nanofibres may be nanorods (solid) or nanotubes (hollow)

Nanostructured materials and associated hierarchy are the subjects of an on-going work item (ISO/TS 80004-4).

A similar approach in promoting a newer term was taken with 1-d objects, where nanoplate was favored for not being widely current in the scientific literature. Other suggestions were considered, but there were frequently secondary associations that the project experts thought should be avoided. This is especially true for the potential that a 1-d nano-object definition might have overlapping connotations with film or coating. In TS27687:2008(E) the note to nano-object addresses this issue by using the wording "discrete nanoscale objects," which aligns with several proposals for the term *nanopowder* (see term 3.16 of PAS 71) and SCENIHR's use of discrete in [7].

Our third question is, "What is a nanomaterial and are there distinguishable categories of nanomaterials?" Published TC 229 standards do not address this point directly, i.e. by offering definitions, but this will change later in 2010 when ISO/TS 80004-1, "**Nanotechnologies – Vocabulary – Part 1: Core terms**" is issued. Consistent with the phraseology from the committee's scope, "*typically, but not exclusively under 100 nm,*" the earlier work items stressed objects below 100 nm in at least 1-d, but this situation will shortly be expanded with the publication of the initial framework and nanomaterial classification documents, TR 12802 and TR 11360. These reports will be informative of JWG1 deliberations. In the meantime, the reader should review PAS 136:2007 for the terms *nanostructured*, *nanomaterial* and *nanoporous*, where they will find many of the issues discussed earlier in the non-SDO section.

It was recognized quite early by TC229 JWG1 experts that a consistent set of definitions would require some categorization of nanotechnologies into the individual nanotechnology fields and a set of core terms provided that both serve to guide the many working groups and avoid repeating issues when new experts joined project teams. JWG1 experts responded by initiating the two work items, "**Nanotechnologies – Terminology – Initial framework model for core concepts**" (ISO/TR 12802) and "**Nanotechnologies – Vocabulary – Part 1: Core terms**" (ISO/TS 80004-1).

The framework document (ISO/TR 12802) addresses several categorizations of nanotechnology: fields of activity, nanomaterials, processes, nanosystems and nanodevices, and properties. An initial listing of 82 pertinent terms were used to populate subject area diagrams expressed as taxonomic hierarchies. The JWG1 experts followed a library science approach based on ANSI/NISO Z39.19-2005 and ISO 2788:1986. Two tests were used to validate each hierarchy: (1) The descriptive "is a" test (a [narrower concept] is a [broader concept]); and (2) The "all-and-some" logic test (Some [broader concepts] are [narrower concepts]. All [narrower concepts] are [broader concepts]). Each of the 12 resulting hierarchies is accompanied by a discussion including advantages and disadvantages. The hierarchies do overlap; in fact, there are three properties frameworks, presaging the commentary found here, and some terms appear in multiple hierarchies. Though not definitive, the framework document certainly provides guidance to future TC229 expert teams.

The core terms document (ISO/TS 80004-1) was balloted in early 2010, and publication is expected in late 2010. This document does provide definitions for *nanotechnology*, *nanomaterial* and *nanostructure*. Nanomaterials may be nano-objects or nanostructured materials, which means that a nanomaterial may be macroscopic in size or nanoscale in size. Increasingly, the concept of nanostructure has gained greater prominence as being the crucial element to this definition. Where nanotechnology may involve control of matter in the sense of precise position control at the nanoscale, a nanostructure is the resulting element that exhibits nanoscale properties or nanoscale phenomena. An isolated nanostructure is a nano-object, and a collection of nanostructures becomes the basis for nanostructured materials.

There are two ISO TC 229 activities in WG4 that involve interactions with JWG1 topics. WG4's remit is setting specifications, which in many respects takes the definitions of JWG1 and uses them in the buyer-seller context. There are no published standards from WG4 that the reader can refer to at this time, but PD 6699-1:2007 is a solid source of the concepts being discussed. Many potential characteristics are identified along with suggested test techniques covering all but three or four of the listed properties. In PD 6699-1:2007, 1-d nano-objects are considered nanoscale films or coatings, which has been mentioned already as a topic of controversy. At this point in time, the WG4 efforts are tending to focus first on differentiating the nanoscale form of a material from the larger scale form. The 60 m^2/g surface area measurement is prominent in these discussions, but it is complemented with direct particle size measurements and TEM pictures for shape. WG4 will likely explore the 19 liaisons with other ISO committees to establish joint working groups before developing specifications affecting those specific applications. It may be necessary in those situations to return to the core term definitions to maintain consistency among TC 229 documents.

The second WG activity is led by CEN/TC 352 under the Vienna agreement. ISO TC 229 national bodies that are not members of CEN have observer status in this consultation process, and the final document is voted on separately by both organizations. PAS 130:2007 was the starting document for that group's discussions. Though not nominally a terminology document, the CEN/TC 352 standard does include concepts such as "*nanoscale phenomenon*" and "*use of the prefix 'nano'*," and it will rely in many respects on the discussions surrounding specifications in WG4. With the recent cosmetics directive [21], some aspects may move from the original voluntary intent to a more mandatory implementation. It may be an oversimplification, but the work of the CEN/TC 352 brings the many SDO and non-SDO issues discussed here into sharper relief. (A further discussion of labeling to be found in Chapter 9 of this book, "Labeling".)

2.2.4 Concluding Terminology Comments

In combining the several sections into one commentary, we observe that there are distinct communities, each with its own view of relevance, each active in creating terms, in defining them and in recommending their interpretation. This confluence of interests affects the definition of nanoscale significantly, as each group collapses its viewpoints into a size-only criterion. The recourse to a size-only criterion arises from the difficulties each group encounters in defining the unique, novel or unexpected properties to be associated with nanoscale materials.

Upon closer examination of the scientific literature, no new phenomena or properties have been noted for nanoscale materials when considered in the broader context of all materials. A simple example would be a surface-mediated catalytic reaction, which is naturally more prominent when a material has a high specific surface area (m^2/g). The ability, however, to control matter with a nearly molecule-by-molecule precision, combined with doping and multi-component compositions, does allow for the amplification of properties that are not normally associated with a specific material in the larger scale. In addition, the excitement associated with nanotechnology has brought attention to our understanding of sub-micron particles and to data gaps surrounding their extrinsic, biological properties.

Three groupings occur within the overall dialog:

1. Those who accept that an upper boundary to *nanoscale* covers a broad spectrum of materials and phenomena with the expectation that a "unique" property is highly likely or that a material's ensemble of properties can be viewed as unique for a specific application
2. Those who accept that some materials will exhibit sharp transitions in a property, while other materials will exhibit gradual changes sufficient to allow for reasonable extrapolation from the large-scale to nanoscale forms
3. Those who primarily approach nanotechnology from an established framework, such as statutory language (a regulatory agency), a mission statement (an NGO, funding agency) or a paradigm used in a neighboring field of study (medicine)

The debate is magnified when discussing materials that have been studied for quite some time. These are usually inorganic having mineralogical names, and that have been either processed (grinding, hence a "top down" description) or synthesized (precipitated, hence a "bottom up" description) for decades. The debate is muted when discussing a newly created, multicomponent material, e.g. encapsulated superparamagnetic iron oxide. The debate's boundary is situated at new atomic structures, such nanotubes for carbon or for ZnO, or with biological molecules of nanoscale dimensions. In the latter three cases, there are rarely larger scale analogs for comparison.

Lastly, there is a significant difficulty in terms of nomenclature. Even in fields such as colloid or catalyst chemistry, there is no nomenclature system to differentiate the several transformations a nanoscale material may undergo throughout the product life cycle. Where the colleagues at E56 have added the concept of informatics to their terminology efforts, the TC229 experts have an exploratory effort in nomenclature, which is the topic for the next section.

2.3 Nomenclature and Nanotechnology

Generally speaking, nomenclature is a formal system that is used to consistently assign recognizable names based on a framework of rules. A good nomenclature system should function like a post office address: the assigned name should allow experts to recognize the nano-object and be a useful means for locating further information. Ideally, such a system should be designed to be able to accommodate the naming of undiscovered entities. The concept of nomenclature enjoys general agreement within the scientific community and represents a systematic means of identification and communication across scientific disciplines, commercial markets, government agencies, and international borders.

Nomenclature systems for chemical substances are grounded in their chemical formulas; familiar chemical formulas are NaCl (salt) and H_2O (water). These simple formulas, when expressed as sodium chloride and dihydrogen oxide, are useful across many technical, governmental, and commercial disciplines to describe the arrangement of the constituent atoms. To describe more complex chemistries, scientists supplement the formulas with additional features such as terms, prefixes, and positional numbering to describe where atoms attach to each other, so that anyone familiar with the language can readily visualize or draw the 3-d structure of the molecule.

2.3.1 Why is Nomenclature Useful in Relation to Standards?

For the research community, a unique name for a specific nano-object would allow for the development of meaningful relationships between nano-objects, their properties and effects. Nomenclature facilitates the repeatability of experimental data among separate research groups, helps support the development and use of standardized reference materials, and serves as a communication tool in grant applications and for the protection of patents.

A specific name assignment for a nano-object would help consumers distinguish it from other products and would strengthen the identification of a substance beyond a trade name for purposes of establishing standards. For example, two manufacturers may use different trade names for their end-use products, but share an ingredient with the same identity. Assigned names will help to foster confidence and broad use of product specifications that are designed around a common understanding and name of the subject ingredient and overall product composition.

Reflecting these practical benefits from nomenclature, standards organizations are finding that there is a need to participate in the development of nomenclature for nanotechnology. While traditional chemical nomenclature rules provide an excellent starting point for naming nano-objects, names that can sufficiently distinguish nano-objects from each other as well as from their larger scale chemical counterparts are generally lacking [27].

The perceived information gap can be illustrated with naming conventions for titanium oxides, useful commercial materials of longstanding that are noteworthy for their wide ranging commercial applications. Titanium dioxide (TiO_2) may have the crystal structure of anatase or rutile depending on the arrangement of titanium and oxygen atoms in the crystal lattice. In the industrial manufacturing process, chemical additives such as aluminum salts are used to promote rutile formation and to lower photocatalytic activity and other additives provide surface treatments to meet end-user performance requirements. In addition to recognized variations in structure, TiO_2 as a category tends to have a distribution of particle sizes as well. Irrespective of whether the particle size distribution is partially within the nanoscale range, entirely within the nanoscale range, or completely outside of the nanoscale range, current chemical nomenclature dictates that the substance be named TiO_2 (with some additional accommodation made to describe crystal structure as noted). When large enough, TiO_2 serves as an excellent white pigment by scattering visible light. When small enough, TiO_2 is transparent to visible light, but absorbs UV radiation. While current chemical nomenclature sufficiently describes the fundamental crystalline structure and the molecular entity, it is insufficient to signal which form of TiO_2 we are referring to, even though they have very different desirable commercial properties. Further differentiation could be accomplished by adding terms to describe porosity or by numbering to indicate a particle size range measurement. At the nanoscale, tubular shapes may also be formed [28], but would currently not be distinguished in their name from non-tubular forms. Because the morphology (shape) of a chemical substance at the nanoscale may have an effect on how the substance performs, it may be desirable for standards setting to differentiate nanoscale titanium dioxide from macro-sized counterparts and from other nanoscale forms due to differing catalytic activities.

In the case of carbon nanotubes, a nomenclature system is lacking beyond citing the number of walls (single, double, multi-walled forms) and the chirality vector. The first investigators tended to distinguish these from conventional forms simply because they were new. Carbon is not new: there are two well-known allotropes of carbon: diamond and graphite [29, 30]. They are characterized by a nominal integer degree of carbon bond hybridization, corresponding to sp^3 tetragonal, sp^2 trigonal, and sp digonal hybridization of the 2s and 2p valence orbitals respectively. IUPAC [31] defines allotropes as “different structural modifications of [an] element,” with allotropic transition considered the “transition of a pure element from one crystal structure to another which contains the same atoms but which has different properties.” Materials that change their crystal structure with external conditions such as temperature and pressure, but where the covalent bonding between the elements remains unchanged, are not true allotropes, but rather polymorphs [32].

The first allotrope of carbon, *diamond* or the isotropic form, consists of tetrahedrally-bonded carbon atoms and typically crystallizes in a face-centered cubic crystal system. The chemical bonding between the carbon atoms is covalent with sp^3 hybridization [33]. However, while the rare diamond polymorph known as lonsdaleite also consists of tetrahedrally-bonded carbon atoms, it crystallizes in a hexagonal crystal system [34–36]. Nonetheless, diamond is represented commonly as *Diamond* with CASRN 7782-40-3 with further distinctions found in the field of minerology.

The second allotrope of carbon, *graphite*, the anisotropic form, consists of layers of hexagonally-arranged, trigonally-bonded carbon atoms in a planar condensed ring system. An individual planar sheet of sp^2-bonded carbons, each atom covalently bound to three neighboring carbon atoms, is known as a *graphene*: these are stacked parallel to each other in layers, connected by weak van der Waals forces. Crystalline allotropic modifications of elements, i.e., polymorphs, are systematically named by adding the Pearson symbol in parenthesis of the name of the atom. This symbol defines the structure of the allotrope in terms of its Bravais lattice (crystal class and type of unit cell) and the number of atoms in its unit cell. Thus, the common form of graphite is carbon (*hP*4), denoting hexagonal primitive – four atoms; and the less common form of graphite is carbon (*hR*6), denoting hexagonal rhombohedral – six atoms [37, 38]. In other words, graphite may be viewed as a finite assembly of graphene units. Both natural and synthetic graphite occur in two crystalline forms with different stacking arrangements, consisting of hexagonal graphite in combination with less than 40% rhombohedral graphite. Further, natural graphite occurs in three principal forms: crystalline flake, lump, and amorphous. Each form exhibits a differentiable suite of physical characteristics. Crystalline flake graphite consists of flat, plate-like particles with angular, rounded, or irregular edges; lump graphite is typically massive and ranges in particle size from extremely fine to coarse; and amorphous graphite is characterized by a low degree of crystallinity and very fine particle size [39]. Graphite spirals are also known [40]. CAS representation of graphite does not distinguish among the two crystalline and several morphological forms: all conform to *graphite* with CASRN 7782-42-5.

In recent years, claims for new, so-called allotropes of carbon have proliferated. For example, fullerenes have been described as "the third form of carbon" after diamond and graphite [41]. Carbon nanotubes have been characterized as a type of fullerene or even claimed as a new carbon allotrope. It has been claimed that such modifications of the primary carbon allotropes may exhibit non-integer or mixed degrees of carbon bond hybridization. [42]. However, in the absence of significant changes in fundamental crystalline structure, as would be demonstrated by geometrical changes in the admantane-like building blocks of diamond or the graphene structure of graphite, new polymorphs of carbon compounds would not necessarily qualify as elemental carbon allotropes.

The term *amorphous carbon* commonly is used to describe carbon materials that do not have any long-range crystalline structure. Short-range order exists, but with deviations of the inter-atomic distances or inter-bonding angles, or both, with respect to the diamond (sp^3 configuration) and graphite (sp^2 configuration) lattices [43]. While amorphous carbon is sometimes cited as an allotrope of carbon, the amorphous carbon of commerce, i.e., coal, soot, and other carbon materials that are neither diamond nor graphite, are not truly amorphous. Rather, these substances consist of polycrystalline diamond or graphite embedded in an amorphous carbon matrix. In accordance with IUPAC nomenclature, which requires that a "sample of an element that has an undefined formula, or is a mixture of allotropes… bear the same name as the atom," amorphous carbon is described as *Carbon* with CASRN 7440-44-0 [44].

2.3.2 Nomenclature Challenges

Telling nano-objects apart by formal names would be desirable because their small size and structure combined with chemical composition may cause nano-objects to behave very differently than larger scale counterparts. There are reports of some materials that do not normally conduct electricity, do so in their nanoscale form. Thus, the chemical names that we currently assign to nanoscale materials (and their underlying chemistry) may not be fully descriptive and leave room for ambiguity or error.

It is recognized and generally accepted that a formal chemical nomenclature will lag behind advances in technology [45]. Such a time lag places standards setting organizations in the field of nanotechnology in a unique role, that of taking steps to see that rules for naming nano-objects keep abreast at the introductory stages of the technology [46].

Encouraging the advancement of tailored nomenclature sooner in time for the nanotechnology field is being attempted in recognition of the role of communication to the success of modern technological advancement. In the absence of a nomenclature system that distinguishes nano-objects from other nanoscale and larger scale counterparts with the same molecular composition, the ability to set standards to measure, characterize, identify, assess, manage or manufacture nano-objects in a reproducible way is presented with a significant challenge.

Equally problematic, in the absence of a definitive set of rules, is a tendency to resort to adding the prefix "nano" to the names of common chemical substances to identify them at the nanoscale, resulting in names such as "nanosilver" or "nanotitanium dioxide." The prefix "nano" has also been used in more general material references such as "nanoparticles," "nanocones," and "carbon nanofibres" (see [4] and ISO/TS 80004-3). Yet, it is equally possible or probable to choose not to use the term "nano" to name objects at the nanoscale, which, in turn, complicates the identification of existing and developmental nano-objects in commercial applications that may be affected by standards setting activities.

For purposes of nanotechnology and standards, a nomenclature system needs to rise to the challenge of providing a precise frame of reference to facilitate product evaluation and commercial development.

2.3.3 Standards Development Organizations and Nomenclature for Nanotechnology

In June 2005, the International Organization for Standardization (ISO) formally established a Technical Committee (ISO/TC 229) to progress standardization in the field of nanotechnology. In 2008, a Nomenclature Task Group was established by Joint Working Group (JWG) 1, Terminology and Nomenclature, and a Task Group Report was finalized in June 2010 at the ISO/TC 229 Plenary session held in Seattle, Washington, USA. The 2009 ISO/TC 229/WG1/TG1 Report on Considerations for Developing Nomenclature Models for Nano-objects defines nomenclature as a system of naming that provides a minimum set of descriptors to identify an object. The TG Report identified ten objectives for an effective nomenclature system for nano-objects that may be used as a basis to guide future work.

In August 2009, a new work item proposal (NWIP) was submitted jointly to ISO/TC 229 by the United States and Canada to prepare a technical report and develop a framework for nomenclature models for nano-objects. ISO/TC 229 approved the NWIP proposal in September 2009 and the first working group session was held at the TC-229 Plenary session in Tel Aviv, Israel in October 2009.

It is TC-229's objective to establish a framework of subclasses of nano-objects that will be used as the basis for developing nomenclature for specific nano-object subclasses. This will include a set of objectives of a nomenclature system, a recommended schedule for developing nomenclature for nano-object subclasses, and discussion of administrative and related challenges.

For this purpose, ISO is collaborating with private organization leaders in the field of chemical nomenclature, including the International Union of Pure and Applied Chemistry (IUPAC), the American Chemical Society (ACS), and the Chemical Abstracts Service (CAS). In this way, methods for supplementing the existing chemical nomenclature systems established and recognized by these nomenclature bodies will be examined to further refine our ability to distinguish nano-objects.

It is hoped that subsequent new work items and associated project groups will evolve for the development of nomenclature models for specific subclasses of nano-objects. The framework exercise is designed to place nanotechnology chemistries into context by indicating the types of materials that are platforms for nanotechnology applications. Such context will provide the international community with a structured view of nanotechnology and facilitate common understanding of nano-objects and their names. Focus will be on nano-objects, namely discrete chemistries with one, two, or three dimensions in the range of approximately 1–100 nm [4].

2.3.4 Overview of Recognized Chemical Nomenclature Bodies

A logical progression for nano-object nomenclature begins with an examination of the basic "workhorse" chemical substances which are emerging as the building blocks for more complex compounds, systems, arrays, and discoveries. Metal oxides and carbon-based substances such as fullerenes and nanotubes are considered a good starting point. At the most fundamental level, these are chemical substances. A "chemical substance" in relevant part may be viewed as any "organic or inorganic substance of a particular molecular identity" [47]. Although there is no ready definition for "particular molecular identity," internationally-accepted chemical nomenclature practice is grounded in the concept that the representation of a particular substance is defined by its molecular composition, which is based on molecular arrangement and bonding structure. Internationally-accepted chemical nomenclature practices are highly relevant to facilitate commercial acceptance by the standards-user community; existing chemical nomenclature systems are thought to be an excellent starting point for discussion.

IUPAC is an organization of technical experts that identify and address needs related to chemical nomenclature for common voluntary usage [48]. The bulk of its chemistry-based nomenclature system distinguishes materials almost exclusively based on molecular composition, and sometimes structure when applicable (such as prefixes for isomers). In the field of nanotechnology, IUPAC has published a nomenclature system for naming fullerenes designed to differentiate between fullerenes based on different atom connectivity [49, 50]. Rules for numbering (C_{60}-I_h)[5,6]fullerene and (C_{70}-$D_{5h(6)}$)[5,6]fullerene were codified in 2002. In 2005, IUPAC issued a supplement containing recommendations for numbering a wide variety of fullerenes of different sizes, with rings of different sizes, from C_{20} to C_{120}, and of various point group symmetries, including low symmetries such as C_s, C_i, and C_1, as well as many fullerenes that have been isolated and well characterized as pristine carbon allotropes or as derivatives. The recommendations are based on the principles established in the earlier publication and aim at the identification of a well-defined, and preferably contiguous helical pathway for numbering. Rules for systematically completing the numbering of fullerene structures for which a contiguous numbering pathway becomes discontiguous are provided by the IUPAC system. It is nevertheless difficult to extend this identical set of rules to other nano-objects because all nano-objects do not exhibit the well-characterized structures of fullerenes, which are viewed by many as molecules.

CAS is a not-for-profit division of the American Chemical Society. CAS has derived a nomenclature system to facilitate its principal business objective of providing information search and retrieval capabilities. It maintains this nomenclature system, closely related to the IUPAC system, for the purpose of database building, abstract preparation, and information retrieval [51]. In particular, CAS sponsors the CAS Registry^{SM}, an authoritative collection of disclosed chemical substance information. CAS offers an arbitrary but unambiguous registry number system to identify the chemical substance. CAS's naming rules will be different, for example, depending on whether the chemical substance has fixed chemical structures (such as discrete chemicals), number of repeating units (such as polymers), or is characterized as a "Unknown or Variable compositions" (UVCB) [52]. One of the advantages of the CAS system is that it is formal when naming a known substance but flexible to accommodate unknown substances by categorizing them as UVCB's. (An example of a UVCB name would be "Chemical A, reaction products with Particle X"). The CAS system is accompanied by a simple, randomly assigned numeric or alpha-numeric identifier for indexing and retrieval. The extent to which a simple index number can be developed into a more complex reference system for obtaining additional information on nanotechnology (also called "smart numbering") is an area that could be explored.

Yet, both IUPAC and CAS nomenclature are based unequivocally on the principle of structure. As noted in the introduction to the CAS Name Selection Manual:

> This manual sets out in detail the entire body of procedures employed by the staff of the Chemical Abstracts Service (CAS) in selecting a unique, reproducible name for every inorganic and organic chemical of defined molecular structure… [53].

The *CAS Name Selection Manual* emphasizes the critical role of structure, rather than physical, chemical, or biological properties, as the basis for CAS index nomenclature, in that:

> A second difference between index nomenclature and commonly used nomenclature is that for the former there must only be one unique name for a structure. Names used by the general chemical public in scientific publications, trade literature of the like, tend to reflect a particular point of interest, such as reactivity and biological activity, rather than similarity in basic structure [54].

In delineating the procedure to determine a CA preferred index name, CAS instructs that "from the structure of the compound," one first determines the highest compound class to which it belongs… on which an index name may be based. In a subsequent step, one should "[n]ame the structural fragments to be cited as substituent prefixes" [55]. Particularly relevant to inorganic carbon compounds, the *CAS Name Section Manual* states that "[t]he names selected for inorganic compounds are based on United States usage, the IUPAC rules… and the representation of chemical structure" [56].

The *IUPAC Red Book*, an internationally-recognized compendium of rules for naming inorganic compounds, states that the "primary aim of chemical nomenclature is simply to provide methodology for assigning descriptors (names and formulae) to chemical substances so that they can be identified without ambiguity, thereby facilitating communication." A nomenclature system "must be recognizable [sic], unambiguous, and general" [57]. Similarly, the *IUPAC Blue Book*, the corresponding compendium of rules for naming organic compounds, states that "[t]o be useful for communication among chemists," chemical nomenclature "should contain within itself an explicit or implied relationship to the structure of the compound, in order that the reader or listener can deduce the structure (and thus the identity) from the name." This purpose requires "a system of principles and rules, the application of which gives rise to a systematic nomenclature" [58]. In describing the functions of chemical nomenclature, the *IUPAC Red Book* states that:

> The first level of nomenclature, beyond totally trivial names, gives some systematic information about a substance, but does not allow the inference of composition… When a name itself allows the inference of the stoichiometric formula of a compound according to general rules, it becomes truly systematic. Only a name at this second level of nomenclature becomes suitable for retrieval purposes. The desire to incorporate information concerning the 3-d structures of substances has grown rapidly, and the systematization of nomenclature has therefore had to expand to a third level of sophistication [59].

Further emphasizing the exclusive role of molecular structure considerations in systematic chemical nomenclature, the *IUPAC Red Book* states that the "systematic naming of an inorganic substance involves the construction of a name from units which are manipulated in accordance with defined procedures to provide compositional and structural information." Appropriate units include "structural, geometric, [and] stereochemical" descriptors [60]. Noteworthy by their complete absence from the IUPAC hierarchical nomenclature scheme are descriptors for physical, chemical, and biological properties.

To summarize the IUPAC and CAS systems, both consist of publicly available rules and guidelines based primarily on molecular composition, and in addition, CAS has numerical identifiers coupled with a searchable information system capable of cataloguing a sizable library of formal chemical names. In light of these fundamental principles of structure-based nomenclature, for reasons previously noted, there is an additional need perceived that these systems can and should be supplemented in real time. It will remain a future possibility whether an authoritative nomenclature body for nanotechnology needs to be established to implement and maintain nomenclature rules and a registry system.

2.3.5 Other Concept

The 2009 paper published by Gentleman and Chan is intended to address the needs of the research community to identify their research materials and for standard test materials [61]. The naming convention suggested by Gentleman and Chan uses physical parameters coupled with a chemical name to distinguish nano-objects from each other and from their larger scale counterparts. The system uses a numerical identifier which points to a specific parameter (e.g., size and shape, core chemistry, ligand, and/or solubility).

2.3.6 Possible Parameters for a Nanotechnology Nomenclature System

Just as with conventional chemical nomenclature, naming rules for nano-objects should be tailored to the needs of the class or sub-class of substances under consideration as determined by experts familiar with these chemistries. To be used and understood, assigned names should not be overly descriptive, complex or lengthy. There are thought to be certain physical-chemical parameters that could be distinguished in a name because they stand out as particularly relevant for nano-objects that share the same chemical composition but exhibit different properties. Two examples are particle size and particle shape.

The particle size of a nano-object can be used distinguish one nano-object from another and from its smaller or larger counterparts. It is probably the easiest to measure and cross-cutting attribute available for this purpose. Its drawbacks include the myriad of methods for measuring particles size that may create "apples to oranges" moments when evaluating the parity between materials. In addition, particle size is a simplistic approach that may offer visualization at the expense of capturing the true scientific characteristics that distinguish the chemistry at work.

Expressing the physical shape of the nano-object before and/or after surface functionalization (such as tubular, spherical, cubic, etc.) permits greater recognition concerning the reactivity and surface area of the substance. For nano-objects such as tubes, the length distribution may be a consideration as well. Nevertheless, the shape of a nano-object is not limited to the simplest geometrical forms, can be rather complex, and may have transitional status. Nano-objects can be composed of a random or periodic arrangements of randomly shaped nanoscale features/structures.

A countervailing consideration for chemical nomenclature is that the use of physical and chemical property distinctions for substances, where there is an analytically-based, detailed chemical composition and a definite chemical structure diagram, would be unprecedented in the rules of structural theory-based CAS and IUPAC nomenclature upon which agencies such as the EPA rely. Parsing out the statutory obligations associated with new chemical determinations in no way diminishes the important role of physical and chemical properties for use in assessing the health and environmental hazard and exposure and risk posed by new (or existing) chemical substances. The risk assessment component of the PMN review into which these considerations are factored, however, is triggered only after EPA determines that the chemical identity is, in fact, not on the Inventory. Much has been made of physical form in particular. On this specific topic, the classic case is long-standing guidance in the United States on silica in which the regulatory agency has repeatedly gone on record that physical form and crystalline structure are fungible (interchangeable) and the former may be disregarded:

> The Agency is aware that silicon dioxide, commonly referred to as silica, occurs and is distributed for commercial purposes in several different physical forms. Inasmuch as the chemical compositions of the various physical forms are the same, EPA does not consider the different physical forms of silica to be separately reportable under TSCA. For the purposes of TSCA, the various physical forms of silica (SiO2) are all considered to be included [62].

The above summarization remains the situation today, with certain limited exceptions. We understand that the EPA, in line with standard chemical and mineralogical practice, treats substances with different crystalline structures as separate chemical substances, but it does not distinguish between substances with the same crystalline structure if they have different physical forms. While all forms of silica have the same molecular formula, (SiO_2), some silicas have different crystalline structures and so must be listed individually on the Inventory. Both CAS and IUPAC emphasize the role of molecular structure considerations in systematic chemical nomenclature. As previously noted, absent from these nomenclature schemes are descriptors for physical, chemical, and biological properties. The question for standards organizations and nomenclature bodies today is whether chemical nomenclature should remain limited to this principle [63].

More traditionally, the identification of reactive function groups is a fundamental piece of information that is communicated through nomenclature. Thus, reactive species are not a new concept to chemical nomenclature. Functionalization is taken to new heights in nanotechnology, however. In many cases, surface-functional aspect are necessary to understand to recognize the true nature of a nano-object. Knowledge of the core chemistry allows for an understanding of the stability of the nano-object when coupled with the surface functionalization. Typical "core" chemistries in nanotechnology are gold, silver, carbon, aluminum oxide and titanium dioxide. The type of bonding to the core is useful information to understanding the substance that may be reasonable to consider in a name, while this is not now commonly done. One nanoparticle core or shell might have a plethora of surface-added species, affecting the nano-objects' properties such as electronic, magnetic, mechanical, surface area, solubility and reactivity. In understanding the surface functionalization of a nano-object, one gains a deeper appreciation of the useful commercial properties as well as its possible degradation products.

Delineated crystal structure is another concept that is not new to chemical nomenclature that has particular utility in the field of nanotechnology. Crystal structure offers specific insight as to the molecular arrangement of a nano-object and may provide insight into its degree of reactivity.

2.3.7 The Distinction Between Characterization and a Name

Recall the analogy of nomenclature to a post office address. If the addressing envelope were also to specify the color of the residence, the number of residents in the home, the property value and tax assessment, and the applicable land use zoning code information, it would be a long address indeed! The data beyond the residence number, street name, city, state and zip code are useful in their own right, but they are not a requirement for reliable delivery of the mail. In addition, having a reliable address permits one, with some additional discrete effort, to locate many additional types of detailed information.

In this same way, a discrete name for a nano-object should not be overly detailed; the expectation should not be that a name will address every performance and behavioral aspect. Instead, it is a reasonable expectation that the name will allow an interested person to locate additional details on the substance.

A discrete name improves the level of confidence that toxicological testing will be performed on the same substance and that such testing will yield reproducible results. Consistent naming rules should allow health and safety professionals to systematically and reliably use information retrieval services to obtain toxicology information indexed by chemical name. Ideally, a name that distinguishes nano-objects with the same chemical composition but different properties improves the ability of these groups to recognize a specific nano-object that presents the potential hazard apart from others. Hazard communication professionals are an important user community for nomenclature. The two endeavors of hazard communication and nomenclature development are distinct from one another, however, and each requires a particular expertise.

Health and safety regulators also are an important user community. For trained regulators, chemical names provide an initial indicator for how the substance might behave in the environment and affect human health and toxicity. Under inventory-based regulatory systems, chemical names provide a communication tool for regulators to signal which substances have undergone government review, which are subject to regulation, or those that require premarket notification.

2.3.8 *Future Nomenclature Directions*

In summary, nomenclature is a formal system used to assign a name to an object based on a framework of rules, enabling the identity of the object to be readily understood. Because of size and chemical interactions with their surroundings, nano-objects may exhibit unexpected properties not seen in their larger counterparts with the same chemical composition. A nomenclature system designed for naming nano-objects would allow the research community, industry, governments and public interest groups to uniquely identify the nano-object is in use, distinguish products from others, protect patents, and communicate effectively across a variety of industries and scientific disciplines.

Since nano-objects may have the same molecular composition as their larger counterparts, adapting and enhancing our existing nomenclature systems for chemicals seems prudent, efficient, and reasonable to promote ease of understanding and widespread use. The existing nomenclature systems for chemicals however, are currently limited in their ability to distinguish chemical substances based on structure and properties other than chemical composition. This has led to current ambiguities in the naming and ready identification of nano-objects.

Ideally, a nomenclature system for nanotechnology should result in names for individual nano-objects that are descriptive enough for a knowledgeable reader to understand key aspects and properties of the object. The naming rules themselves should be simple and clear enough so that different users will be able to generate the same results.

Any concurrent attempt to develop and name scientific discoveries creates a need for cooperation and information sharing among standards organizations and chemical nomenclature experts. Standards organizations are helping to set the pace for incorporating the latest information and best practices in commercial applications of nanotechnology. The rules governing the nomenclature system for nano-objects will need to be based on what is known, acknowledge current limitations and minimize uncertainty, and accept that the "correct" properties or parameters for naming will continue to advance. Given that all of the parameters for identifying various nano-objects may not be known for some time, the risk that engaging in this exercise is premature should be accompanied by the commitment to develop a system able to withstand rigorous re-examination and the ability to adjust to new information. However, there are communication and knowledge benefits to beginning the development process now with due care as described, providing that there remains the understanding of the need for possible near term or longer term course corrections as experience is gained in deploying such a system.

2.4 Final Remarks

The current activities by long-standing standards developing organizations or other interested institutions have been reviewed through the perspective of terminology, vocabulary, controlled vocabulary and nomenclature. In addition to offering an overview, this chapter identified challenges and opportunities, as well as the broader societal issues surrounding prudent nanotechnology regulation. There are many organizations involved in this effort, and though emphasis was placed on the two major standards developing organizations, ISO and ASTM International, the complexity of the issues and the high level of interest surrounding nanotechnology will continue to attract active, global participation. And, there are, of course, daily announcements on new developments arising from the considerable investment being made in global nanotechnology research.

The reader is encouraged to become an active participant. Unlike the more established standards development topics, nanotechnology is definitely in flux and likely to undergo dramatic changes in direction and understanding. The effort to find the proper balance between pragmatic and rigorous definitions and definitive and flexible taxonomies or to decide when a topic area is sufficiently established to propose definitions are in themselves important determinants of this field's success. This is especially true for those aspects influencing regulatory decisions on the prudent introduction of products arising from nanotechnology.

References

1. House of Lords Science and Technology Committee: Nanotechnologies and Food, vol. I and II. http://www.publications.parliament.uk/pa/ld200910/ldselect/ldsctech/22/22i.pdf (2010). Accessed 16 Feb 2010
2. Kuhn, T.S.: The Structure of Scientific Revolutions, 2nd edn. The University of Chicago Press, Chicago (1970). Enlarged
3. Webster's New International Dictionary, Unabridged. G. & C. Merriam Co., Springfield (1954)
4. International Organization of Standardization: Nanotechnologies – Terminology and Definitions for Nano-Objects, ISO/TS 27687:2008(E). ISO, Geneva, Switzerland (2008)
5. ASTM International: E 2456-06 Terminology for Nanotechnology. ASTM International, West Conshohocken, USA (2008)
6. The Royal Society and The Royal Academy of Engineering: Nanoscience and nanotechnologies: opportunities and uncertainties. The Royal Society, London. http://www.nanotec.org.uk/finalReport.htm (2004). Accessed 16 Feb 2010
7. EC Scientific Committee on Emerging and Newly Identified Health Risks (SCENIHR) to the European Commission: Opinion on the scientific aspects of the existing and proposed definitions relating to products of nanoscience and nanotechnologies. http://www.ec.europa.eu/health/ph_risk/committees/04_scenihr/docs/scenihr_o_012.pdf (2007). Accessed 16 Feb 2010
8. EC Scientific Committee on Consumer Products to the European Commission: Opinion on safety of nanomaterials in consumer products. http://www.ec.europa.eu/health/ph_risk/committees/04_sccp/docs/sccp_o_123.pdf (2007). Accessed 16 Feb 2010
9. ETC Group: A tiny primer on nano-scale technologies and The Little Bang Theory. ETC Group. http://www.etcgroup.org/upload/publication/55/01/tinyprimer_english.pdf. Accessed 15 Feb 2010
10. Hette, A.: Nanotechnology small matter, many unknowns. Swiss Reinsurance Company. http://www.swissre.com/pws/research%20publications/risk%20and%20expertise/risk%20perception/nanotechnology_small_matter_many_unknowns_pdf_page.html (2004). Accessed 16 Feb 2010
11. Soil Association: Nanotechnologies and food evidence to House of Lords Science and Technology Select Committee. http://www.parliament.uk/documents/upload/st136SoilAssociation.pdf (2009). Accessed 15 Feb 2010

12. Department for Environment Food and Rural Affairs: UK voluntary reporting scheme for engineered nanoscale materials. http://www.defra.gov.uk/environment/quality/nanotech/documents/vrs-nanoscale.pdf (2006). Accessed 16 Feb 2010
13. Breggin, L., Falkner, R., Jaspers, N., Pendergrass, J., Porter, R.: Securing the promise of nanotechnologies – towards transatlantic cooperation. Chatham House (The Royal Institute of International Affairs), London. http://www.elistore.org/reports_detail.asp?ID=11116 (2009). Accessed 16 Feb 2010
14. Miller, G., Senjen, R.: Out of the laboratory and on to our plates, 2nd edn. Friends of the Earth, Australia. http://www.foe.org/out-laboratory-and-our-plates (2008). Accessed 16 Feb 2010
15. Swiss Federal Office of Public Health and Swiss Federal Office for the Environment: Precautionary matrix for synthetic nanomaterials, version 1.0. http://www.bag.admin.ch/themen/chemikalien/00228/00510/05626/index.html?lang=en (2008). Accessed 16 Feb 2010
16. Auffan, M., Rose, J., Bottero, J.-Y., Lowry, G.V., Jolivet, J.-P., Wiesner, M.R.: Towards a definition of inorganic nanoparticles from an environmental, health and safety perspective. Nat. Nanotechnol. **4**, 634–641 (2009)
17. Powell, J.J., Thoree, V., Pele, L.C.: Dietary microparticles and their impact on tolerance and immune responsiveness of the gastrointestinal tract. Br. J. Nutr. **98**(Suppl. 1), S59–S63 (2007)
18. Ostrowski, A.D., Martin, T., Conti, J., Hurt, I., Herr Harthorn, B.: Nanotoxicology: characterizing the scientific literature, 2000–2007. J. Nanopart. Res. **11**, 251–257 (2009)
19. Sanguansri, P., Augustin, M.A.: Nanoscale materials development – a food industry perspective. Trends Food Sci. Technol. **17**, 547–556 (2006)
20. Chemical Selection Working Group, U.S. Food & Drug Administration: Nanoscale materials [no specified CAS]; nomination and review of toxicological literature. http://www.ntp.niehs.nih.gov/ntp/htdocs/Chem_Background/ExSumPdf/Nanoscale_materials.pdf (2006). Accessed 16 Feb 2010
21. caNanoLab glossary. https://wiki.nci.nih.gov/display/ICR/caNanoLab. Accessed 20 Feb 2010
22. Cosmetics directive. http://www.eur-lex.europa.eu/LexUriServ/LexUriServ.do?uri=OJ:L:2009:342:0059:0209:en:PDF. Accessed 16 Feb 2010
23. No. Cion 5431/08 DENLEG 6 CODEC 59: proposal for a regulation of the European parliament and of the council on novel foods and amending regulation. http://www.register.consilium.europa.eu/pdf/en/09/st10/st10754-ad01.en09.pdf. Accessed 17 Feb 2010
24. EPA Scientific Panel (Heringa et alia): Evaluation of the hazard and exosure associated with nanosilver and other nanometal pesticide products, SAP minutes no. 2010-01. http://www.epa.gov/scipoly/sap/meetings/2009/november/110309ameetingminutes.pdf (2010). Accessed 16 Feb 2010
25. EFSA Scientific Panel (Barlow et alia): The potential risks arising from nanoscience and nanotechnologies on food and feed safety. EFSA J. **958**, 1–39. http://www.efsa.europa.eu/en/scdocs/doc/sc_op_ej958_nano_en,3.pdf (2009). Accessed 16 Feb 2010
26. TC 229 scope statement. http://www.iso.org/iso/iso_technical_committee?commid=381983. Accessed 18 Feb 2010
27. TSCA inventory status of nanoscale substances – general approach (USEPA). http://www.epa.gov/oppt/nano/nmsp-inventorypaper.pdf. Accessed Mar 2010
28. Yang, D., Qi, L., Ma, J.: Eggshell membrane templating of hierarchically ordered macroporous networks composed of TiO_2 tubes. Adv. Mater. **14**, 1543–1546 (2002)
29. Long, J.C., Criscione, J.M.: Carbon survey. In: Kirk-Othmer Encyclopedia of Chemical Technology, vol. 4, p. 733. Wiley, New York (2003)
30. Lagow, R.J.: Synthesis of linear acetylenic carbon: the 'sp' carbon allotrope. Science **267**, 362–367 (1995)
31. IUPAC: Compendium of Chemical Terminology (the "Gold Book"), 2nd edn, Compiled by McNaught, A.D., Wilkinson, A. Blackwell, Oxford. http://www.goldbook.iupac.org/ (1997). Accessed Mar 2010
32. IUPAC red book at IR-11.7 ("polymorphism") and the "IUPAC compendium of chemical terminology". http://www.iupac.org/publications/compendium/index.html (updated, online version of Compendium of Chemical Terminology, 2nd edn. Blackwell, 1990)

33. IUPAC gold book, "diamond"
34. Rode, A., Gamaly, E.G., Christy, A.G., Fitz Gerald, J.G., Hyde, S.T., Elliman, R.G., Luther-Davies, B., Veinger, A.I., Androulakis, J., Giapintzakis, J.: Unconventional magnetism in all-carbon nanofoam. Phys. Rev. B **70**, 054407 (2004)
35. Shigley, J.: Diamond, natural. In: Kirk-Othmer Encyclopedia of Chemical Technology, vol. 8, p. 519. Wiley, Hoboken (2002)
36. Wentorf Jr., H.: Diamond, synthetic. In: Kirk-Othmer Encyclopedia of Chemical Technology, vol. 8, p. 530. Wiley, Hoboken (1992)
37. IUPAC gold book, "graphite," "graphene layer," and "rhombohedral graphite"
38. IUPAC red book 2004 at IR-3.5.3 (crystalline allotropic modifications of elements)
39. Kalyoncu, R.S., Taylor Jr., H.A.: Natural graphite. In: Kirk-Othmer Encyclopedia of Chemical Technology, vol. 12, p. 771. Wiley, Hoboken (2002)
40. Horn, F.H.: Spiral growth on graphite. Nature **170**, 581 (1952)
41. Taylor, R.: Fullerenes. In: Kirk-Othmer Encyclopedia of Chemical Technology, vol. 12, p. 228. Wiley, Hoboken (2002)
42. Leshchev, D.V., Kozyrev, S.V.: Grouping of carbon clusters and new structures. Fullerenes Nanotubes Carbon Nanostruct. **14**, 533–536 (2006)
43. IUPAC gold book, "amorphous carbon"
44. IUPAC red book 2004 at IR-3.4.1 (name of an element of infinite or indefinite molecular formula or structure)
45. Crane, E.J.: Chemical nomenclature in the United States. In: Chemical Nomenclature: A Collection of Papers Comprising the Symposium on Chemical Nomenclature Presented at the Diamond Jubilee of the American Chemical Society, September 1951, vol. 8, pp. 55–64. American Chemical Society, Washington (1953)
46. ISO TC-229, JWG 1, PG 11: Nomenclature framework project for nano-objects (2009)
47. Section 3(2) (A) of the Act (15 U.S.C. §2602(2)(A))
48. International Union of Pure and Applied Chemistry (IUPAC). http://www.iupac.org/. Accessed Mar 2010
49. Powell, W.H., Cozzi, F., Moss, G.P., Thilgen, C., Hwu, R.J.-R., Yerin, A.: Nomenclature for the C60-Ih and C70–D5h(6) fullerenes. Pure Appl. Chem. **74**, 629–695 (2002)
50. Cozzi, F., Powell, W.H., Thilgen, C.: Numbering of fullerenes (IUPAC recommendations 2005). Pure Appl. Chem. **77**, 843–923 (2005)
51. Chemical Abstracts Service: About CAS. http://www.cas.org/
52. Chemical Abstracts Service: CAS registry and CAS registry numbers. http://www.cas.org/
53. Introduction: Chemical Abstract Services Chemical Name Selection Manual, vol. I. American Chemical Society, Washington (1982)
54. Principles of general index nomenclature: CAS Name Selection Manual at A-005, vol. I
55. CAS Name Selection Manual at A-006, vol. I
56. CAS Name Selection Manual, vol. III at IN-1
57. Nomenclature of inorganic chemistry, provisional recommendations 2004 at IR-1.3
58. Preamble: A Guide to IUPAC Nomenclature of Organic Compounds (Recommendations). Blackwell (1993) (Blue Book)
59. IUPAC red book 2004 at IR-1.4 (functions of chemical nomenclature)
60. IUPAC Red Book 2004 at IR-1.5.2 (name construction)
61. Gentleman, D., Chan, W.: A systematic nomenclature for codifying engineered nanostructures. Small **5**, 426–431 (2009)
62. Letter from Henry Lau to John Lewinson, Degussa Corporation, Dec 21, 1990 (IC-3070); Letter from Henry P. Lau, EPA, to Daniel C. Hakes, 3M (Nov 19, 1993) (IC-4482)
63. Sellers, K.: Nanoscale materials: definition and properties. In: Sellers, K., Mackay, C., Bergeson, L.L., Clough, S.R., Hoyt, M., Chen, J., Henry, K., Hamblen, J. (eds.) Nanotechnology and the Environment. CRC, Boca Raton (2009)

Chapter 3
Nanoscale Reference Materials

Gert Roebben, Hendrik Emons, and Georg Reiners

3.1 Introduction

3.1.1 The Growing Use of Reference Materials

Globalisation of both science and trade has increased the relevance of the comparability of measurement data whether in research, industry or regulatory contexts. Reference materials (RMs) are essential tools in the quest for comparable and reliable measurement results, a quest which laboratories, worldwide, are tasked with every day. An explicit acknowledgement of the importance of RMs in today's measurement systems is found, for instance, in the laboratory accreditation standards, such as ISO/IEC 17025 [1].

The awareness of the need for reliable RMs is growing in parallel with the increase in the number of laboratories operating under formal accreditation systems. As a result, the demand, therefore the production, hence the variety, and also the use of RMs, are all increasing and they are expected to further increase in the years to come. Figure 3.1 shows the evolution of the number of logins and search results in the COMAR database (an international database for certified RMs; see also Sect. 5.2). Even if the observed increase is also a result of an increased number of materials in the database and the increased awareness of the existence of the database, the trend nicely illustrates the growing attention for RMs over the last 7 years.

G. Roebben (✉)
Institute for Reference Materials and Measurements,
Joint Research Centre of the European Commission, Geel, Belgium
e-mail: gert.roebben@ec.europa.eu

V. Murashov and J. Howard (eds.), *Nanotechnology Standards*,
Nanostructure Science and Technology, DOI 10.1007/978-1-4419-7853-0_3,

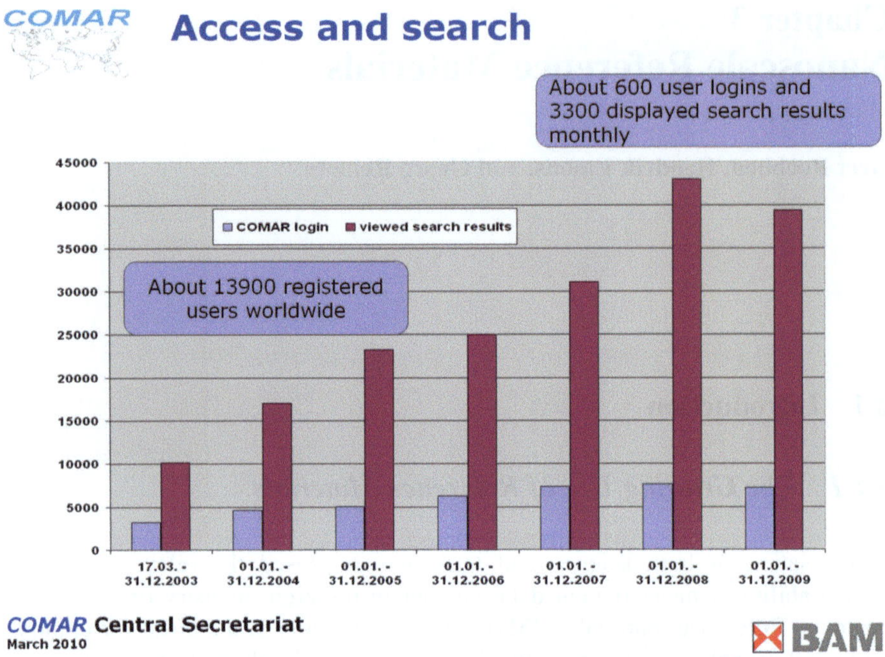

Fig. 3.1 Evolution of the access and search statistics for the COMAR database, an international database for certified RMs, maintained by the Federal Institute for Materials Research and Testing, Berlin, Germany (BAM) [2]

3.1.2 The Term "Nanoscale"

Many organisations, worldwide, have developed or are developing a terminology for use in the field of nanotechnology. (An overview is presented in Chap. 2 by Abe et al. in this book, and in a recent report of the European Commission's Joint Research Centre (JRC) [3].) Where possible, this chapter will use the terminology for nanotechnologies developed by the International Organization for Standardization (ISO), as well as the ISO terminology for reference materials and for metrology in general. In this Chapter, where definitions of ISO terms are quoted, these have been reproduced with the permission of the International Organization for Standardization (ISO). The corresponding ISO documents are referenced and can be purchased from the website of the ISO central secretariat (http://www.iso.org/isostore). Copyright remains with ISO.

The ISO terminology for nanotechnology is based on the crucial term *nanoscale*, which is defined as the size range from approximately 1 nm–100 nm [4]. A nanoparticle, for example, is a particle with all three external dimensions at the nanoscale [4]. ISO has recently released a document containing additional definitions for a number of nanotechnology core terms. For example, the term nanomaterial will be used as a collective term for nano-objects (particulate materials with one, two or three external dimensions at the *nanoscale* [4]) and nanostructured materials (materials with internal features at the *nanoscale*). The definitions of these terms are available in the on-line, searchable ISO Concept database [5].

3.1.3 Nanotechnology Needs Reference Materials

The common characteristic of all nanotechnology-specific measurement questions is the scale issue. Either it concerns measurements that have to be done with a nanoscale spatial resolution, or with an affinity for the nanoscale features of nanomaterials. It is exactly in this new, challenging measurement domain that a significant number of methods need to be developed and validated. It is expected that the reliability of the results of these new methods, and those from existing methods pushed beyond their previous detection or quantification limits, will increasingly be challenged. This is exactly why nanoscale RMs are needed, as is for example highlighted in the December 2007 Strategic Plan of the (USA) National Nanotechnology Initiative [6].

3.1.4 Structure of the Chapter

This chapter will try to elucidate the field of nanoscale RMs. To prepare the reader for nanoscale RMs, first a number of key concepts and terms from existing documentary standards that pertain to the production and use of RMs are introduced (Sect. 3.2). Then the critical issues, specific to nanoscale RMs, are highlighted (Sect. 3.3), and a number of typical examples is given (Sect. 3.4). The chapter will end with an outlook on the current developments and trends (Sect. 3.5).

3.2 Generic Issues in Reference Materials Production and Use

3.2.1 The Role of ISO/REMCO

RMs have been used for many centuries (think of the weight or length standards applied already in ancient cultures). Today RMs are used in all fields of natural sciences, from physics to chemistry and biology, and for many different purposes, from method development, calibration and validation to laboratory internal quality control or external proficiency testing. In each of these disciplines and for most of these purposes, a particular RM terminology has been developed. It was not until relatively recently that the conceptual similarities of RMs across scientific fields and application sectors have been explored, recognised and more systematically investigated.

The codification of the developed consensus has, to a large extent, been the work of the ISO Committee on Reference Materials (ISO/REMCO) which was created in 1975.

The terms of reference of ISO/REMCO include such tasks as to establish definitions, concepts and classification of RMs, to determine the basic characteristics of RMs in dependence on their use, to formulate criteria for the selection of publications referenced in ISO documents (covering also legal aspects), to prepare guidelines for technical committees when dealing with RM issues in ISO documents, and to propose, as far as necessary, actions to be taken on RM issues required for ISO work. So far ISO/REMCO has produced and revised six ISO Guides and one Technical Report. ISO/REMCO is currently developing several new work items [7].

In 1992, definitions for terms used in connection with RMs were first proposed, and recently, the corresponding ISO Guide 30 has been amended [8] with newly agreed definitions for the terms reference material (RM) and certified reference material (CRM). Carefully analysing both definitions, the major characteristics of RMs are revealed in the following section.

3.2.2 Reference Material

The definition of reference material is [8]:

> (A RM is) a material, sufficiently *homogeneous* and *stable* with respect to one or more specified properties, which has been established to be fit for its intended use in a measurement process.

Notes:

1. RM is a generic term.
2. Properties can be quantitative or qualitative, e.g. identity of substances or species.
3. Uses may include the calibration of a measurement system, assessment of a measurement procedure, assigning values to other materials, and quality control.
4. A single RM cannot be used for both calibration and validation of results in the same measurement procedure.
5. International Vocabulary of Metrology – Basic and General Concepts and Associated Terms has an analogous definition (VIM – ISO/IEC Guide 99:2007 [9] 5.13), but restricts the term "measurement" to apply to quantitative values and not to qualitative properties. However, Note 3 of ISO/IEC Guide 99:2007, 5.13, specifically includes the concept of qualitative attributes, called "nominal properties".

3.2.2.1 Homogeneity

The first characteristic mentioned in the definition is that of homogeneity. It is of obvious importance that the value of the specified property (this can be any property, from chemical composition, to density, to particle size, to thermal conductivity and further), assigned to the RM, can be measured on any *part* of the

RM on which the RM user can or is allowed to measure. *Part* of the RM can be "one of the 2,000 ampoules" of which the particular batch of particles suspension RM exists, but it can also be "any of the thousands of lines" on the step height standard for calibration of a scanning probe microscope.

With the possible exception of a number of gas mixtures or ideal solutions, the homogeneity of a RM is never perfect: there will always be minor differences between samples or within sub-samples of the RM. The cost associated with the processing of RMs is closely related to the desired or required between- and within-sample homogeneity. Since it is such an important RM attribute, the remaining heterogeneity must be experimentally assessed, to demonstrate that it is small enough for the RM to be fit for its intended use. Typically, this demonstration consists of measuring, under repeatability conditions, a random selection of "parts" (sub-samples, ampoules, areas,...) of the RM and the calculation of the standard deviation of the obtained results [10]. A contribution corresponding to the detected (or the maximum undetectable) heterogeneity must be included in the uncertainty budget of the property value assigned to the RM.

The homogeneity criterion is inherently linked to the choice or definition of a "minimum sample volume" for the RM. This can be the minimum number of steps on the step height standard for which the measured step heights must be averaged, or the volume of a nanoparticle suspension to be injected in a centrifuge for calibration. With decreasing sample volume, it is expected to see larger variations of the measured property values between samples. Therefore, a meaningful value for the RM homogeneity requires the statement of the corresponding minimum sample volume.

3.2.2.2 Stability

The second main RM characteristic is the RM's stability, or, more precisely, the constancy of the value assigned for the property of interest. Unavoidably, there will be a distance both between the place and time of production and the place and time of use of the RM. During the transport and during storage periods (shelf-life) between production and use, the property values assigned to the RM shall not change beyond a level that is pre-defined as acceptable. The stability of the RM can be demonstrated, for example, by performing an isochronous study [11], which essentially consists of a series of measurements on samples which had been pre-exposed to a scheme of temperature excursions, which mimic the extremes of the realistically expectable storage and transport conditions.

3.2.2.3 Notes to the RM Definition

The RM definition comes with five Notes, two of which are explained here.

"NOTE 1: RM is a generic term."

It is a matter of fact, and not necessarily a problem, that in many fields other terms than "RM" are used to denote something that essentially is the same thing. It must however be stressed that, whenever a material is used for any of the purposes described in this chapter, it must meet the minimum characteristics captured in the above definition of "RM". The specific material can have additional characteristics, such as "coming with particular information" (e.g. a certificate of analysis), "be a metal" or "shall only be used as a blind sample for proficiency testing within the next 6 weeks". RM is therefore considered as a generic term, a common name for a large family of materials [12].

"NOTE 2: Uses may include the calibration of a measurement system, assessment of a measurement procedure, assigning values to other materials, and quality control."

In Sect. 3.4 the different RM usages are explained in more detail, and in Sect. 3.5, several examples of nanoscale RMs will be given, to illustrate these different RM usages.

3.2.3 Certified Reference Material

The definition of certified reference material is [8]:

> (A CRM is) a reference material characterized by a *metrologically valid procedure* for one or more specified properties, accompanied by a *certificate* that provides the value of the specified property, its associated *uncertainty*, and a statement of *metrological traceability*.

3.2.3.1 Metrologically Valid Characterisation

The term "metrologically valid characterisation" is clarified in NOTE 2 to the definition, which states that "metrologically valid procedures for the production and certification of RMs are given in, among others, ISO Guides 34 and 35" [13, 14]. The principles of metrological validity are essentially the requirement to have metrologically traceable certified values with properly estimated uncertainties. Moreover, a sufficient confirmation of the measured property is required, which can be achieved by using different methods – where possible – and the exclusion of the human factor in the measurement process as far as possible.

3.2.3.2 Certificate

A CRM is necessarily accompanied by a certificate, which holds the information that is essential to the use of the CRM. ISO Guide 31 [15] gives guidance on the content of certificates, which should include, among other, sections on the identification of the CRM producer and on the certified value and its uncertainty, a traceability statement, an expiration date, a minimum amount of sample to be used, and the method(s) used for certification, instructions for use and storage.

3.2.3.3 Uncertainty of the Certified Value

The uncertainty of the certified value is needed to make a meaningful comparison between the certified value and a value measured on the CRM in the user's lab [16]. The uncertainty of the certified value is typically a combined uncertainty, containing contributions from the homogeneity assessment, the stability evaluation, and the characterisation measurements (the measurements performed to determine the certified value).

The efforts of a CRM producer are aimed at obtaining an uncertainty of the certified value which is as small as possible, or at least smaller than a pre-defined acceptable value. The smaller the uncertainty, the more powerful the CRM is when searching for method bias. Also, if the CRM is used for the calibration of a method, then the uncertainty of the certified value is a direct contribution to the overall method uncertainty.

3.2.3.4 Metrological Traceability

The formal definition of the term metrological traceability is found in the International Vocabulary of Metrology (VIM) [9]:

> Property of a measurement result whereby the result can be related to a reference through a documented unbroken chain of calibrations, each contributing to the measurement uncertainty.

Note 1 to this definition states:

> …, a "reference" can be a definition of a measurement unit through its practical realization, or a measurement procedure including the measurement unit for a non-ordinal quantity, or a measurement standard…

The practical realisation of metrological traceability is often a challenge. But in essence, the concept is relatively simple: metrological traceability is the answer to the question: "With which measurement results can I compare the value I measured?" Measurement results and certified values, but also legally defined threshold values or target values in industrial production processes, can only be compared with one another if the stated or measured values are traceable to the same reference.

3.2.4 *Different Usage of Certified and Non-certified RMs*

3.2.4.1 Accuracy, Trueness and Precision

A major difference between certified and non-certified RMs is related to the terms precision and trueness, which are the two main components of the accuracy of a measurement result. Precision is related to the statistical variation of repeated measurements. A method is precise if it produces highly repeatable measurement results. A method gives true results if it produces, on average, a value which is correct (or *without bias*). Precision-related measurement issues can be checked with any sufficiently homogeneous and stable material, which are the basic characteristics of any RM, also the non-certified RMs. Trueness of a method can only be checked with a CRM, as this comes with a certified value, which is a best estimate of the true value.

3.2.4.2 Calibration

Many instruments need to be calibrated to establish the relationship between measured signal and the property to be measured. Materials for calibration obviously need to have reliably assigned values, hence need to fulfil the requirements of CRMs. There are two fundamentally different applications of CRMs in the calibration process. The first are materials used to calibrate instrument or method parameters (calibration of wavelengths, mass, temperature etc.), whereas the second group is used for generating calibration curves of the measurand (property intended to be measured) versus the basic instrument response.

3.2.4.3 Method Validation

If a laboratory wants to validate one of the methods it intends to use, then a number of validation issues have to be addressed. Among these issues are repeatability and intermediation precision, which require tests on series of samples under specific conditions. The variability detected during these tests is used to assess the corresponding contributions to the overall measurement uncertainty of the results produced with the method. Obviously, one should avoid heterogeneity between the samples used in the method validation study as such heterogeneity contributes to the variability under repeatability or intermediate precision conditions. Therefore the use of a homogeneous and stable set of samples is recommended, i.e. the use of samples of a RM.

Another issue for method validation is "trueness" assessment. This ideally is based on the comparison of a test result obtained by the lab using its method on a sample with known property value and corresponding uncertainty, hence with the qualities of a CRM. If the obtained measurement result corresponds with the certified value, taking into account the combined measurement uncertainty and the uncertainty of the certified value, then the absence of bias can be concluded and the method can be considered to provide true values [16].

3.2.4.4 Statistical Quality Control

Statistical quality control consists of periodic assessments or qualifications of the proper functioning of an instrument. Results of statistical quality control tests can, for example, be represented in quality control charts, which intend to visualise the variability over time of the performance of an instrument or method. Again, as in the previous section on method validation, it is desirable to eliminate as much as possible the variability due to the test samples from the variability in the periodically obtained measurement results. Materials for this purpose have to be homogeneous and stable, hence must fulfil the requirements of RMs.

3.2.4.5 Interlaboratory Comparisons

Also materials used for interlaboratory comparisons (such as laboratory proficiency tests or studies of interlaboratory method reproducibility as part of a method validation study) must be homogeneous and stable (at least for the duration of the test). They therefore also need to fulfil all requirements of RMs.

3.3 Critical Issues Related to Nanoscale RMs

Nanotechnology is, essentially, about the development and use of structures, components and materials at the nanoscale. While nanotechnology is a relatively new term, the desire to design and work, and therefore also to measure at smaller scales, is a classical endeavour, common to all major scientific disciplines. Therefore, "nanotechnology" covers a broad field of (potential) applications (each may be complemented by the prefix "nano"): electronics, optics, pharmacy, medical technology, mechanical engineering and others. In each of these fields, new materials have been produced, some of which have external dimensions in the nano-range ("nano-objects"), such as nanotubes or nanoparticles, others have internal, structural features at the nanoscale ("nanostructured materials"), such as the multilayered thin film structures that are the basis of modern electronic devices.

The mentioned application areas each have different demands in terms of critical parameters to be measured (for example geometrical, optical, electrical, magnetic and others), and correspondingly, the demands for RMs are different. This section describes the generic, critical issues for RMs for nanotechnology-related measurements. In Sect. 3.4, examples will be given to illustrate these generic issues and to reveal a number of more specific issues.

3.3.1 Definition of the "Measurand"

For many nanoscale measurands, no "reference methods" (sometimes called "primary methods"), or more correctly "reference measurement procedures," [9] exist. Therefore, the certified values of the RMs have to be obtained using the methods at the same metrological level as those for which their application is planned. This has also consequences for the achievable quality characteristics of the certified values, as the method performance will often not be superior for the characterization of the RM in comparison to routine applications.

This lack of "higher order methods" is not unique to nanotechnology. It is in fact frequently encountered in various, if not all, fields of science: chemistry, biology, and (material) physics, and is essentially related to the issue of method-defined properties [17]. A method-defined property is a property which is not intrinsic to the test object, but which is to a certain extent defined by the measurement procedure. Naturally, it is (most often) not possible to compare the value of a method-defined property with a value obtained with another method. It is equally obvious that a measurement result obtained for such a property is only meaningful when a correct reference is made to the employed measurement procedure.

From a traceability and comparability point-of-view, method-defined properties are not the most desirable. Without any doubt, the advancement of science will allow a better understanding of material properties and of the test methods to assess these properties. This should result in an increasing number of methods which assess materials properties in a method-independent way. However, it has to be acknowledged that method-defined properties are often of practical use and of industrial or regulatory significance, and deserve consideration from the metrological perspective, including the provision of RMs.

3.3.2 Traceability Statements

The default aim in metrology is to achieve SI-traceable measurement results: measurement results that are traceable to the International Systems of Units, which consists of the units kilogram (for mass), metre (for length), second (for time), candela (for luminous intensity), ampere (for electric current), kelvin (for thermodynamic temperature), and mole (for amount of substance). It is therefore most common that RMs come with property values that are SI-traceable. However, in the case of nanomaterial characterisation, with a majority of the (current) test methods delivering procedural or method-defined property values, it is not straightforward to achieve SI-traceability. This is an issue for other measurement fields as well, and therefore the partners of the European Reference Materials (ERM®) cooperation (BAM, the JRC Institute for Reference Materials and Measurements (IRMM) and LGC) have developed a dedicated policy for the traceability statements on the certificates of their RMs [18]. The policy is developed to provide an answer to questions such as "Can a result be traceable to a method if the quantity value can be linked to the SI?" or "Can a result be traceable to the SI if the measurand depends on the applied measurement method?". The ERM® answer to these questions is that it is a matter of more precision in the traceability statement. The ERM® policy distinguishes, within a traceability statement, the issues of "identity" (or definition of the measurand) and "quantity value" (the number and its unit). The message is that the quantity value can be traceable to the SI system, also when the measurand is operationally defined. However, this requires that the operationally defined nature of the measurand is spelled out with the reported measurement result (or certified value), and that all influence parameters, that affect the measurand, were measured or calibrated in an SI-traceable manner [19].

3.3.3 Laboratory Qualification

The production of RMs critically depends on the availability of laboratories that are proficient in the measurement of the properties for which values need to be assigned or certified. In the best of cases, CRM producers can rely on measurements performed by formally accredited laboratories (including their own labs). However, only few laboratories have an accreditation scope which includes measurements at the nanoscale. This implies that the CRM producer has to establish for himself whether candidate laboratories adopt at least the ISO/IEC 17025 approach, and whether these laboratories can demonstrate their competence, for example with results from inter-laboratory comparisons. Only a limited number of proficiency tests have been organised for measurements on nanoparticles [20, 21]. In practice this means that RM certification projects have to be preceded by a preliminary inter-laboratory study, using a non-certified RM, to establish a sufficiently large group of laboratories with demonstrated expertise [22].

3.3.4 Homogeneity and Stability

In Sect. 3.2, the two fundamental characteristics of a RM were explained: homogeneity and stability. These RM properties are particularly challenging to be realised and proven in the nano-context.

3.3.4.1 Homogeneity at the Nanoscale

The homogeneity of an RM, as was mentioned earlier, is directly related to the defined minimum sample volume. In the case of nanoscale measurements, obviously the typical sample volumes are orders of magnitude smaller than those encountered in classical, macro-analysis methods. This implies that the averaging effect on which RM producers can "rely" when producing RMs for macroscale analysis does not, or to a much lesser extent, play its role for many nanoscale RMs.

A distinction can be made here between the characterisation of nanomaterials, in general, and the characterisation of materials at the nanoscale, which are two different issues. Many nanomaterials or components containing nano-objects (e.g. nanocomposites) will also have to be characterized for their macroscopic properties (e.g. electrical conductance, optical properties, mechanical strength, and toughness). There is no immediate need for specific nanoscale RMs for the quality assurance of these macroscopic measurement methods, unless, of course, the property value levels of nanomaterials are beyond the range of property values for non-nanomaterials. In the latter case, it may be necessary to develop nanomaterial RMs, for example to enable calibration of the method at the extremes of its measurement range.

3.3.4.2 Stability of Nanoscale Structural Features

The "stability" of the RM property for which a certified or assigned value was determined, is related to the stability of the material's microstructure. It is well-known that nanosized or nanostructured materials have the tendency to agglomerate or coarsen, as the result of the natural tendency to minimise surface energy. Also, one needs to make sure that the property of the sample is not changed by the measurement process. For example, electron beam irradiation can change the structure either by heating the sample, etching the sample or contaminating the sample.

An obvious example to illustrate the problem of stability of the properties of nanomaterial RMs is that of a powder with certified particle size. It is virtually impossible to avoid the agglomeration of dry nanoparticles. This is why the vast majority of RMs for the calibration or verification of nanoparticle size measurement instruments consists of stabilised suspensions. In these suspensions, agglomeration is eliminated or reduced through the interfacial properties, especially due to surface charges that develop on the suspended nanoparticles, which renders them mutually repulsive and the suspension stable.

3.4 Examples of RMs for Nanotechnology

3.4.1 Areas of Application for Nanoscale Reference Materials

The umbrella term "nanomaterial" collects a large variety of materials, which have only one thing in common: the characteristic properties of these nanomaterials are related to their external size or their internal nanoscale structure. Several schemes are being developed to classify nanomaterials for a certain purpose. With respect to the use of nanomaterials as RMs, they can be classified according to the measurement methods for which they are developed:

1. RMs for methods characterizing **nano-objects**: number of nano-objects (for example concentration of nanoparticles on a solid surface), their size (and size distribution), morphology (e.g. aspect ratio) or chemical composition (including surface chemical composition, and functionalisation);
2. RMs for methods characterizing **thin surface coatings/films** and interfaces: flatness, step height, film thickness, roughness & topography (characterization of

moth-eye structures, e.g. for solar cells), 3D-structures, indentation hardness, Young's Modulus, chemical composition (depth profiles, functionalisation, and sharpness of interface);

3. RMs for methods characterizing **surface nanostructures or masks**: width and height of strips, periodic steps, precision of structural/geometrical "repeating units", pattern dimensions, critical dimensions, and 3D-structures;
4. RMs for methods characterizing **nanoporous materials**, filter, catalysts: porosity, size distribution of pores, distribution of pores in a solid, and (specific) surface area;
5. RMs for methods characterizing **solid nanostructured materials**: crystal size, dispersion homogeneity, and wear resistance.

3.4.2 A Database of Existing Nanoscale RMs

A number of organisations have created databases of available RMs. Most databases are specific for one area (such as GeoReM, the Max Planck Institute database for RMs of geological and environmental interest, or the database of the Joint Committee on Traceability in Laboratory Medicine (JCTLM) for higher-order RMs for the field of in vitro diagnostics). A more generic database was created by COMAR, a non-commercial network of national and international organisations, created as a spin-off of the ISO/REMCO activities, which is open for further international participation. COMAR has created an international, common database of available CRMs. The COMAR database is hosted on the BAM website [2], and is fed via a global system of national contact points, who can upload new and edit existing entries in the database. The COMAR database is restricted to CRMs, and does not take up non-certified RMs. Given the early stages of development of many of the new nanoscale measurement methods (see Sect. 3.3), the number of nanoscale CRMs is limited.

Yet, quite a number of non-certified nanoscale RMs have been developed and are becoming increasingly available. Originated as an idea in the working group "Measurement and Characterisation" of ISO/TC 229 "Nanotechnologies", and based on an initial list of "standards for the calibration of instruments for dimensional nanometrology" [23], members of the German ISO/TC 229 delegation have created a freely accessible database of nanoscale RMs. World-wide commercially available nanoscale RMs are catalogued in an on-line accessible database [24]. Currently (April 2010), 65 entries (15 of which are CRMs) from 19 providers are listed in 13 categories (flatness, film thickness, single steps, periodic steps, step gratings, lateral (X-Y-axis, 1-dim), lateral (X-Y-axis, 2-dim), critical dimensions, 3-dim, nano-objects, nanocrystalline materials, porosity, depth profiling resolution). The certified quantities of the RMs cover the range between 0.3 nm up to 1,000 nm. Fifty RMs have assigned size values below 100 nm. For each RM a pdf data sheet can be downloaded. The collected information comprises name and description, type of RM, RM category, certified quantities and units, test methods that can be calibrated with the (C)RM, characterization methods used, applications, and the provider (web links).

The CRMs included in the database are usually offered by the national (or transnational) metrology institutes (NMIs) and their designated institutes, such as the Physikalisch-Technische Bundesanstalt (PTB), BAM, IRMM, the National Institute for Standards and Technology (NIST), the National Institute of Advanced Industrial Science and Technology (AIST), etc.. CRMs can also be produced by non-NMIs, such as commercial companies or research institutes. Unfortunately, the data necessary from a metrological point of view are often incomplete (e.g. quantitative details on uncertainty and/or the information on the used measurement methods are not always given by the providers). Currently these non-compliant (C)RMs are included in the database, but as the number of available (C)RMs increases, and the database is updated, a more selective attitude can be adopted. In the following paragraphs some examples of nanoscale RMs with different status (RM vs. CRM) are reviewed.

3.4.3 RMs for Nanoparticle Size Analysis

Over the years, the range of available RMs for particle size analysis has naturally extended to smaller particle sizes. Well-known are the polystyrene (PS) latex RMs, which are available from several CRM producers. The particles in the PS and polyvinylchloride (PVC) latex materials are highly spherical and can be made highly monodisperse, and provide excellent calibration tools for a number of methods.

Recently, particle RMs with sub-50 nm assigned values were released by NIST in its series of colloidal gold RMs (RM 8011-8012-8013 [25]). These RMs consist of citrate-stabilized gold nanoparticles in dilute suspension. Several methods were used for the characterisation of these materials, resulting in a list of method-defined values for atomic force microscopy (AFM), scanning electron microscopy (SEM), transmission electron microscopy (TEM), differential mobility analysis (DMA), dynamic light scattering (DLS), and small angle x-ray scattering (SAXS). The nominal reference values are 10 nm, 30 nm and 60 nm, and the assigned uncertainties are typically around 1 nm (relative expanded uncertainties up from 1% and, for SAXS, up to 20%). Any of the methods divergence issues are currently being studied.

The reference values assigned to the Au colloids are a best estimate of the true value provided by NIST where all known or suspected sources of bias have not been fully investigated by NIST. This is why the materials are RMs and not CRMs. The trade-off between an RM and CRM is the investment of time versus improved functionality. It is important to get RMs into the industry; an RM can be more rapidly produced than a CRM. The RM 8011-8012-8013 materials were intended primarily to evaluate and qualify methodology and/or instrument performance related to the physical/dimensional characterization of nanoscale particles used in pre-clinical biomedical research. The RM may also be useful in the development and evaluation of in vitro assays designed to assess the biological response (e.g. cytotoxity, hemolysis) of nanomaterials, and for use in inter-laboratory test comparisons.

Similarly, IRMM-304 [26] is also an RM consisting of an aqueous suspension of nanoparticles, albeit not a gold colloid. The silica nanoparticles in IRMM-304 have assigned values for their hydrodynamic diameter, as do the NIST RMs 8011-8012-8013. The IRMM-304 also includes the Stokes diameter. More in particular, the method-defined properties are $\overline{x}_{DLS}$ (the intensity-weighted harmonic mean diameter, as obtained via the cumulants method and via frequency analysis, methods described in ISO 13321 [27] and ISO 22412 [28]) and $x_{St,m}$ (the modal Stokes' diameter, as obtained via the centrifugal liquid sedimentation method described in ISO 13318-2 [29]). The nominal equivalent sphere diameters are 40 nm with relative expanded uncertainties between 5% and 10%.

The use of IRMM-304, an RM, is limited to quality control issues such as method development, proficiency tests, or control charting [30]. Since the values assigned to IRMM-304 correspond with method-defined properties, the certification of the values can not profit from measurements with primary methods in one or a few expert or reference laboratories. In such case, ISO Guide 35 [14] recommends to pass via an interlaboratory comparison of a larger number of expert laboratories, to reduce the impact of operator- or laboratory-specific factors in the measurement process. IRMM is currently running such an international interlaboratory comparison (ILC), which is intended to deliver updated $\overline{x}_{DLS}$ and $x_{St,m}$ values with increased metrological reliability, allowing the certification of the property values of IRMM-304. It is pointed out here that there is a difference between an interlaboratory study of the reproducibility of a method, using an RM, and an interlaboratory study with the aim of determining the property value of the candidate CRM. In the former ILC, participants are not necessarily expert laboratories. In the latter ILC, participants must be qualified prior to their selection as a collaborator in the certification process, ideally based on a formal accreditation for the relevant measurement method.

3.4.4 RMs for Measurement of Film Thickness

While the term "film thickness" seems to leave little room for interpretation, in practice film thickness is expressed in different ways. The most obvious way is to measure and express film thickness as a length, with a value (potentially) traceable to the SI unit *metre*. The alternative is to express the film thickness as the *areal density of atoms or molecules* (the number of atoms or molecules per unit surface area) which are deposited or implanted in the film or surface layer.

The NMIJ CRM 5202-a is an example of the former, length case. This CRM consists of a SiO_2/Si multilayer structure grown using a radio-frequency magnetron sputtering method on a Si substrate [31]. The certified quantity is the thickness of 4 of the individual films (nominal mean thickness: 20 nm, relative expanded uncertainty: 3%). The layer thicknesses are certified in units of length via grazing incidence X-ray reflectometry. The CRM can be applied to control the precision of analysis and to regulate the measurement conditions in depth profile analysis by ion sputtering (used with Auger Electron Spectroscopy (AES), Electron Spectroscopy for Chemical Analysis (ESCA), and Secondary Ion Mass Spectrometry (SIMS)).

BCR-261 is an example of the other, areal density case. It is a CRM consisting of tantalum pentoxide film on tantalum foil (nominal thickness values 30 nm and 100 nm [32]). The oxide layers are grown by anodic oxidation evenly on both sides of the foils. The BCR-261 certified property is the areal density of oxygen atoms. For the 30 nm film, the mean areal density of oxygen atoms is 1.72×10^{21} m^{-2} (relative expanded uncertainty = 4%), for the 100 nm film, the mean areal density of oxygen atoms is 5.40×10^{21} m^{-2} (relative expanded uncertainty = 2%). The certified property is obtained via nuclear reaction analysis, elastic recoil detection analysis and Rutherford backscattering spectrometry, methods which do not measure the "dimensional" thickness of a layer, but a number or a fraction of atoms or isotopes of a particular element. BCR-261 is intended for use in calibration of various surface analysis methods employing different techniques and equipment. It can also be used for assessing and optimizing the depth resolving capability and sputtering yield of surface analysis instruments, as it has been assigned additional, non-certified, information values (for interfacial resolution and sputtering yield).

3.4.5 RMs for Chemical Contrast Imaging

BAM-L200 is a CRM consisting of a nanoscale stripe pattern for testing of lateral resolution and calibration of length scale, prepared by epitaxial growth resulting in layers with sharp interfaces [33]. Multilayer systems of AlGaAs and GaAs are well known in optoelectronics and their preparation with metalorganic vapour phase epitaxy (MOVPE) is a well established technology. Due to the good fit of the lattice constants of GaAs and AlAs it is possible to prepare thick layers (several hundreds of nanometers) with a great difference in elemental composition. For this reason the system GaAs – Al0.7Ga0.3As was used. Additionally some thin layers of In0.2Ga0.8As are included in the layer stack. The partial substitution of Ga by Al gives a sufficient material contrast for all tested methods of surface analysis. The layer stack has a total thickness of about 12 μm, and the certified values of the individual lines in the pattern range from 3.5 nm to more than 4,000 nm. The certified values are obtained with a calibrated TEM. The results of length measurements on TEM images can be made SI-traceable: the calibration of the scale of the TEM images can be done via measurements of the crystal lattice spacings for a calibrant material, whose lattice spacings have been determined via, for example, diffractometry. TEM is the imaging method with the highest lateral resolution and gives maximum sharpness of the images and correspondingly maximum accuracy in length measurement.

BAM-L200 can be used for all methods of surface analysis and surface imaging techniques which are sensitive to the material contrast between Al0.7Ga0.3As and GaAs. Successful tests have been accomplished with SIMS, AES, Energy Dispersive X-ray Spectrometry (EDS) and ESCA.

3.4.6 RMs for Measurement of Surface Topography

Surface topography at the nanoscale is a field that developed thanks to the invention and development of scanning probe microscopy (SPM). Whereas traditional electron microscopy does provide access to 2D information, the added value and the strong point of SPM is in its sensitivity for out-of-plane features and dimensions. A large number and variety of step height reference materials currently exist for the calibration of these out-of-plane dimensional measurements.

For example, for the 3D calibration of AFMs and optical interferometric microscopes the VLSI Surface Topography Standard [34], a combination of step height and pitch, can be used. This RM has a pitch cluster patterned in a layer of silicon dioxide. The pitch cluster contains three distinct grid patterns consisting of arrays of alternating bars and spaces with extremely uniform pitch in both the X and Y direction. The RM's have pitches between 1.8 μm and 20 μm. The vertical step heights are 18 nm, 44 nm, 100 nm or 180 nm. The topographic patterns are very regular, allowing accurate measurement across the entire working area of the standard. The expanded uncertainty of the nominally 18 nm certified step heights is about 5%.

Another, recently released RM that can be used for the calibration of the in-plane dimensional measurements of scanning probe microscopes is the NIST RM 8820 [35]. This RM consists of a pattern of pitch structures produced with 193 nm ultraviolet light lithography. Since the pattern is not only topographical in nature but also chemical (Si vs. SiO_2) NIST RM 8820 can also be used for the calibration of the lateral scale of electron and particle beam instruments. The smallest pitch has a nominal value of 200 nm, and the corresponding expanded uncertainty is around 5%.

More special cases are:

1. The MMC-40, which is a RM consisting of a 3D pyramid with 520 nano-markers [36], produced by an automated focused ion beam (FIB) patterning process. The certified property is the step height, which nominally is 600 nm, and which is certified as the height difference at the position of different "nano"-markers. MMC-40 is intended for one-step 3D Calibration of SEM and AFM instruments. Although the calibration lengths are above nanoscale it is a tool with relevance for nanoscale analytical methods.
2. The PA01 porous aluminum test structure [37], which consist of a thin film of hexagonal, open pores (cells). Since the thickness of the partitions between the pores is about 5 nm, the radius of the spikes formed at the intersections of the partitions is only 2 nm (approximately). The test structure is therefore well suited to test the performance and the shape of AFM tips (see Fig. 3.2).

3.4.7 RMs for Surface Area Measurements

One of the recurring explanations for the difference in behaviour between nanostructured materials and regular materials is the difference in surface area.

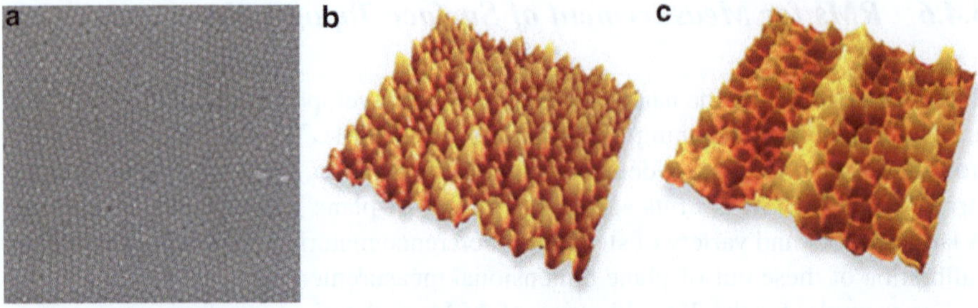

Fig. 3.2 SEM (**a**) and AFM (**b**, **c**) images [38] of the PA01 porous aluminum foil. The AFM image (**c**) has been obtained with an AFM tip with a lower tip radius than the AFM image (**b**) (Images: courtesy Mikromasch, Tallin, Estonia)

BAM-P108 is a CRM consisting of activated nanoporous carbon with large BET surface [39]. One might argue whether the nanoporous carbon is an engineered nanomaterial, but the value of the material's certified BET specific surface area (550 m^2/g, uncertainty = 5 m^2/g) is certainly in the range of relevance for particulate nanomaterials. The BET specific surface area was measured with the static volumetric method using nitrogen at 77.3 K. The CRM is used for calibration and checking of instruments used for the determination of the BET specific surface area by this method. In the same field, NIST is preparing the release of SRM 1898, which consists of an aggregated powder of TiO_2 nanoparticles.

3.4.8 RMs for Powder Porosity Measurements

ERM-FD107 is an example of a porosity CRM, and consists of a microporous zeolite powder (Faujasite type) [40]. The certified quantities are the specific micropore volume (0.217 cm^3/g, uncertainty = 0.002 cm^3/g) and the median pore width (0.86 nm, relative uncertainty = 0.02%). The analytical method used for certification was gas adsorption (nitrogen) at 77.3 K. The material is used in the calibration of methods measuring specific micropore volume and median pore width. It was originally certified as BAM-PM107, and was one of the first materials to be accepted as European Reference Material (ERM®) in 2004.

3.4.9 RMs for Carbon Nanotube Characterisation

Carbon nanotubes (CNTs) constitute a prime example of what are called high-aspect-ratio-nanoparticles (HARNs). These materials derive special properties and application possibilities from their special shape which also has resulted in health and safety concerns. Unfortunately, some of the earlier reports on CNT toxicity seem to have been flawed by effects stemming from non-CNT fractions in the tested CNT materials, such as the heavy metals remaining as an impurity after having served as catalyst in the CNT production process. Also, there is a large variety of CNT materials, with single wall (SWCNT) or multiwall (MWCNT) versions, functionalised or filled, agglomerated or dispersed. It is clear that an improved understanding of the CNT properties requires the development of purified CNT materials for use in physico-chemical and biological testing to better isolate the effects that are specific for the nanostructure of the CNT materials. However, real-world CNTs materials are likely to contain contaminants such as residual transition metal catalysts and other carbon-based reaction by-products.

A particular effort in the development of the desired test and RMs is done at NIST, where several CNT related materials are currently being developed, including a raw soot material and purified "Bucky Paper" (both certified for elemental composition), as well as a purified, length-sorted SWCNT material. The latter RM will be available in three different length fractions, enabling systematic studies of the effect of CNT length on their properties and behaviour.

3.5 Current Developments and Trends for the Future

3.5.1 Scientific Challenges for Nanoscale RMs

There are several scientific challenges associated with the further development of nanoscale RMs:

1. There is a clear need to better identify and describe in sufficient detail the relevant properties for which RMs are needed. Without such detailed description, it is not possible to achieve the required comparability of measurements that is needed, for example, in regulatory issues.
2. There is a need to design and realize sustainable metrological traceability chains/nets that enable the determination of certified values. Especially for those methods that produce method-defined or procedural values, the possible concepts and approaches are still under discussion. The issue of method-defined properties is not unique for nanotechnology-related measurements; it is a challenge also for other fields such as materials characterisation [17]. It is therefore important to actively seek common and generic solutions for this issue.
3. There is a need to improve and newly develop and validate sufficiently accurate measurement procedures (including the evaluation of measurement uncertainties). Initial reports on method validations and uncertainty budgets do point to issues where progress could be made [41].
4. There is a need to define and internationally harmonize reference methods for crucial functional properties of nanomaterials. This is a task that has to be taken up by international standard development organisations, such as ISO, IEC, and other.

3.5.2 Laboratory Accreditation and Regulation

The scientific issues mentioned in the previous section are essential prerequisites for the production of RMs, and even more for the production of CRMs. Combining the significant research efforts and technology investments with the moderate sales numbers of CRMs inevitably results in a high unit price for CRMs. It can be difficult to justify the purchase of a costly CRM. Obvious reasons for the purchase of RMs are related to the quality assurance requirements imposed on accredited laboratories. Often these laboratories operate under such formal accreditation scheme in order to meet the eligibility requirements associated with measurements that have to be run for regulatory purposes. If the field of nanotechnology will become the subject of specific regulatory initiatives, then these are expected to lead to specific measurement requirements.

3.5.3 Collaboration

The worldwide expressed need for RMs largely exceeds the current RM production capacities. In order to increase their combined efficiency and quality, a number of European CRM producers has organised themselves in the ERM® consortium [42]. In this consortium, LGC, BAM and IRMM collaborate intensively. Similar initiatives have started in Asia. In addition to that, bilateral collaboration agreements between metrology institutes often comprise a mutual commitment to support each others RM development, for example by participation in the characterisation studies of the RMs.

Glossary

AES	Auger electron spectrometry
AFM	Atomic force microscope
BAM	Bundesanstalt für Materialforschung und –prüfung
BET	Brunauer-Emmett-Teller (inventors of the BET technique for surface area measurements)
CNT	Carbon nanotube
CRM	Certified reference material
ERM	European Reference Materials
ESCA	Electron spectroscopy chemical analysis
IEC	International Electrotechnical Commission
ILC	Interlaboratory comparison
IRMM	Institute for Reference Materials and Measurements
ISO	International Organization for Standardization
ISO/REMCO	The ISO Committee on Reference Materials

JRC	Joint Research Centre of the European Commission
NIST	National Institute for Standards and Technology (USA)
NMI	National Metrology Institute
NMIJ	National Metrology Institute of Japan
PS	Polystyrene
RM	Reference material
SAXS	Small-angle x-ray scattering
SEM	Scanning electron microscope
SI	International System of Units
SIMS	Secondary ion mass spectrometry
SPM	Scanning probe microscope
SWCNT	Single wall carbon nanotube
TC	Technical Committee
TEM	Transmission electron microscope
TS	Technical Specification
USA	United States of America
VIM	International Vocabulary of Metrology

References

1. International Organization for Standardization: ISO/IEC 17025:2005 General Requirements for the Competence of Testing and Calibration Laboratories. ISO, Geneva (2005)
2. Steiger, Th., Pradel, R.: COMAR Secretariat. http://www.comar.bam.de (2010)
3. Lövestam, G., Rauscher, H., Roebben, G., Sokull Klütgen, B., Gibson, N., Putaud, J-Ph, Stamm, H.: Considerations on a Definition of Nanomaterial for Regulatory Purposes. Publications Office of the European Union, Luxembourg (2010). ISBN 978-92-79-16014-1
4. International Organization for Standardization: ISO/TS 27687:2008 Nanotechnologies – Terminology and Definitions for Nano-Objects – Nanoparticle, Nanofibre and Nanoplate. ISO, Geneva (2008)
5. http://cdb.iso.org
6. The National Nanotechnology Coordination Office: The National Nanotechnology Initiative Strategic Plan December 2007. Subcommittee on Nanoscale Science, Engineering, and Technology, Committee on Technology, National Science and Technology Council, The National Nanotechnology Coordination Office, Washington, DC (2007)
7. http://www.iso.org/iso/standards_development/technical_committees/other_bodies/iso_technical_committee.htm?commid=55002
8. International Organization for Standardization: ISO Guide 30:1992/Amd 1:2008, Revision of Definitions for Reference Material and Certified Reference Material. ISO, Geneva (2008)
9. International Organization for Standardization: ISO/IEC Guide 99:2007, International Vocabulary of Metrology – Basic and General Concepts and Associated Terms (VIM). ISO, Geneva (2007)
10. Linsinger, T.P.J., Pauwels, J., van der Veen, A.M.H., Schimmel, H., Lamberty, A.: Homogeneity and stability of reference materials. Accred. Qual. Assur. **6**, 20–25 (2001)
11. Lamberty, A., Schimmel, H., Pauwels, J.: The study of the stability of reference materials by isochronous measurements. Fresenius J Anal Chem **361**, 359–361 (1998)
12. Emons, H.: The 'RM family' – Identification of all of its members. Accred. Qual. Assur. **10**, 690–691 (2006)

13. International Organization for Standardization: ISO Guide 34: Reference Materials – General Requirements for the Competence of Reference Material Producers. ISO, Geneva (2009)
14. International Organization for Standardization: ISO Guide 35: Reference Materials – General and Statistical Principles for Certification. ISO, Geneva (2006)
15. International Organization for Standardization: ISO Guide 31:2000, Reference materials – Contents of Certificates and Labels. ISO, Geneva (2000)
16. Linsinger, T.: ERM Application Note 1, Comparison of a measurement result with the certified value. European Reference Materials. http://www.erc-crm.org (2005)
17. Roebben, G., Linsinger, T.P.J., Lamberty, A., Emons, H.: Metrological traceability of the measured values of properties of engineering materials. Metrologia **47**, S23–S31 (2010)
18. Emons, H.: Policy for the statement of metrological traceability on certificates of ERM® certified reference materials. European Reference Materials. http://www.erm-crm.org (2008)
19. Koeber, R., Linsinger, T., Emons, H.: An approach for more precise statements of metrological traceability on reference material certificates. Accred. Qual. Assur. **15**, 255–262 (2010)
20. Wang, C.Y., Fu, W.E., Lin, H.L., Peng, G.S.: Preliminary study on nanoparticle sizes under the APEC technology cooperative framework. Meas. Sci. Technol. **18**, 487–495 (2007)
21. ASTM Committee E56 on Nanotechnology: Interlaboratory Study to Establish Precision Statements for ASTM E2490-09 Standard Guide for Measurement of Particle Size Distribution of Nanomaterials in Suspension by Photon Correlation Spectroscopy (PCS). Research Report E56-1001, ASTM Committee E56 on Nanotechnology, Subcommittee E56.02 on Characterization: Physical, Chemical, and Toxicological Properties, April 2009
22. Lamberty, A., Franks, K., Braun, A., Kestens, V., Roebben, G., Linsinger, T.: Interlaboratory comparison of methods for the measurement of particle size, effective particle density and zeta potential of silica nanoparticles in an aqueous solution. JRC Scientific and Technical Reports, IRMM Internal Report RM-10-003, 2010
23. Koenders, L., Dziomba, T., Thomson-Schmidt, P., Wilkening, G.: Standards for the calibration of instruments for dimensional nanometrology. In: Wilkening, G., Koenders, L. (eds.) Nanoscale Calibration Standards and Methods: Dimensional and Related Measurements in the Micro- and Nanometer Range, pp. 245–258. Wiley-VCH, Weinheim, Germany (2005). ISBN 3-527-40502-X
24. http://www.nano-refmat.bam.de
25. https://www-s.nist.gov/srmors/view_detail.cfm?srm=8011
26. https://irmm.jrc.ec.europa.eu/rmcatalogue/detailsrmcatalogue.do?referenceMaterial=I-0304
27. International Organization for Standardization: ISO 13321:1996, Particle Size Analysis – Photon Correlation Spectroscopy. ISO, Geneva (1996)
28. International Organization for Standardization: ISO 22412:2008, Particle Size Analysis – Dynamic Light Scattering (DLS). ISO, Geneva (2008)
29. International Organization for Standardization: ISO 13318–2:2007, Determination of Particle Size Distribution by Centrifugal Liquid Sedimentation Methods – Part 2: Photocentrifuge Method. ISO, Geneva (2007)
30. Kestens, V., Braun, A., Couteau, O., Franks, K., Lamberty, A., Linsinger, T., Roebben, G.: The use of a colloidal silica reference material IRMM-304 for quality control in nanoparticle sizing by dynamic light scattering and centrifugal sedimentation. Presented at the 6th world congress on particle technology, WCPT6, April 2010
31. http://www.nmij.jp/english/service/C/crm/2_E.pdf
32. https://irmm.jrc.ec.europa.eu/rmcatalogue/detailsrmcatalogue.do?referenceMaterial=0261T
33. https://www.webshop.bam.de/product_info.php?cPath=2282_2315&products_id=3225&PHPSESSID=qgckabxnhs
34. http://www.vlsistandards.com/products/dimensional/ststandards.asp?sid=47
35. https://rproxy.nist.gov/srmors/view_detail.cfm?srm=8820
36. http://www.m2c-calibration.com/index.php?top=2&lang=2
37. http://www.spmtips.com/pa
38. Cheng Sui, Y., Saniger, J.M.: Characterization of anodic porous alumina by AFM. Mater. Lett. **48**, 127–136 (2001)

39. https://www.webshop.bam.de/product_info.php?cPath=2282_2304_2305&products_id=3673&PHPSESSID=1557431f1e94a62fecb4c015498a94f1
40. http://www.erm-crm.org/ermcrmCatalogue/list.do
41. Braun, A., Kestens, V., Franks, K., Couteau, O., Lamberty, A., Linsinger, T., Roebben, G.: Validation of dynamic light scattering and differential centrifugal sedimentation methods for nanoparticles characterisation. Presented at the 6th world congress on particle technology, WCPT6, April 2010
42. http://www.erm-crm.org

Chapter 4
Nanoscale Metrology and Needs for an Emerging Technology

Jennifer E. Decker and Alan G. Steele

4.1 Introduction

Nanotechnology is generally defined as the study, exploitation and/or manipulation of matter with size range from approximately 1 to 100 nm. The focus of nanotechnology is largely on the new and novel properties and/or functionalities of traditional substances when they have structures of nano-scale dimensions. Science continues to push frontiers of knowledge, and the transformation of science into technology is underpinned by profound understanding and predictive models, which can only be attained via measurement results which are widely reliable and comparable. Therefore, measurement science and metrology are essential for nano-scale manufacturing of new materials, devices and products. Metrology, the science of measurement, differs from measurement itself. Attributes of metrology include: quantitative knowledge, traceability, evaluation of measurement uncertainty, repeatability, and reproducibility. Measurement traceability is defined [1] as the "*property of a measurement result whereby the result can be related to a reference through a documented unbroken chain of calibrations, each contributing to the measurement uncertainty.*" Traceability to one single reference implies that results obtained by different measurement techniques can be compared with each other on a common scale. At the present time, the International System of Units ("SI") provides a framework for such a global reference.

Measurements that are traceable to international standards are stepping stones to reliable characterization and evaluation of materials – shape, size, and properties such as hardness, stability, hydrophilicity/hydrophobicity, and melting temperature – enabling manufacturers to achieve credible and reproducible results. This, in turn, leads to quality products with widest possible acceptance in the global marketplace. Moreover, uncovering and enhancing the understanding of new properties that lead to imaginative value-added applications relies heavily

J.E. Decker (✉)
Institute for National Measurement Standards, National Research Council of Canada,
Building M-36, 1200 Montreal Road, Ottawa, ON K1A 0R6, Canada
e-mail: Jennifer.Decker@nrc-cnrc.gc.ca

V. Murashov and J. Howard (eds.), *Nanotechnology Standards*,
Nanostructure Science and Technology, DOI 10.1007/978-1-4419-7853-0_4,

on measuring tools that can provide repeatable and reproducible measurement results which are interchangeable with other manufacturers world-wide. Manufacturers possessing the best measuring tools and metrology "know-how" have a leading advantage in developing superior innovative technologies.

Measurement capabilities at the national metrology institutes (NMIs) are maintained at the state-of-the-art so as to support commercialization of new technologies and requirements for international trade. Trade agreements demand demonstrated equivalence between the measurement standards and accreditation systems of buyer and seller nations and therefore metrology is vital to the regulation of trade, the resolution of trade disputes and the reduction of technical barriers to trade.[1]

An important precursor to trading nanotechnology products is the demonstration of product safety as regards human health and the environment. Reliable detection and characterization of nanomaterials is imperative for our understanding in the study of their toxicological behaviours and to draw conclusions that attain global agreement. Metrological tools and techniques are required for experimental demonstration of safety prior to commercial application.

Nanotechnology poses a unique challenge in this regard because many of the tools currently used in the laboratory are difficult to transfer to the shop floor for many reasons – sophistication of equipment, and required expertise of the operator, for example. Tools currently used in the lab need to be modified and/or new ones developed for industry application.

4.2 International Cooperation

Given the broad scope of nanotechnology, the international community is increasingly combining efforts on metrology to develop and promote good metrological practices in the new areas of measurement particular to nanotechnologies. The NMIs cooperate at the international level by way of the Consultative Committees (CCs) of the International Committee for Weights and Measures (CIPM) and the joint ISO/TC229 and IEC/TC113 Working Groups. The NMIs work together towards establishing primary measurement standards for nanotechnology, and increasingly towards development of documentary standards with metrological content. NMIs cooperate via international measurement comparison studies to develop and harmonize accurate measurement techniques and calibrations, thereby establishing internationally-recognized client services and measurement capabilities. Cooperation and partnerships accelerate accumulation of knowledge and leverage resources towards the rapid development, and international acceptance, of the measurement capabilities required for nanotechnology products and services.

[1] World Trade Organization (WTO) and aspects of Technical Barriers to trade: http://www.wto.org/english/thewto_e/whatis_e/tif_e/agrm4_e.htm#TRS

Moreover, harmonization of primary measurement techniques and capabilities supports establishment of regulations on nanotechnology products for environment, health, safety, and trade, which are in the early stages of discussion. Another important advantage of collaboration exists with the potential to serve clients in the interim by quickly setting in place relevant calibration services and providing time to consider future capital investments. Knowledge building and sharing is of primary importance in establishing the scientific basis for mutual recognition of capabilities and calibration services amongst economies. The CIPM Mutual Recognition Arrangement (MRA) [2] allows for recognition of the equivalence of the services offered by the NMIs. It is based on the results of international comparisons and validation exercises which provide the confidence that measurements are indeed as equivalent as we think they are.

Applications of nanotechnologies could be fast-tracked by implementation of metrological thinking along the entire hierarchy of measurement and documentary standards from R&D proof of concept to the "shop floor". Nevertheless, developing a general global plan for nanotechnologies is a challenge because individual economies have varied priorities and the development of nanotechnologies is somewhat of a moving target *vis a vis* industry; however, maintaining communication and active participation in international activities such as standards and metrology development and management go a long way towards ensuring harmonization and cooperation as nanotechnology commercialization develops. Individual labs and funding agencies can then be aware and augment development of measurement science projects in nanotechnologies aimed at topical areas of particular interest to clients in environmental and occupational health and safety (EHS), toxicology and commercialization.

Communication and cooperation amongst very different areas of science and technology is on the rise – for example toxicologists and metrologists have met together to discuss, understand and identify gaps and needs for moving nanoscience forwards. The series of NNI workshops [3] focused on several application areas important to fundamental science, health, safety and regulatory issues. On the international scene, there is also increasing communication, coordination and collaboration as evidenced by the recent BIPM Workshop on Nanoscale Metrology [4] where attendance consisted of half metrologists and half from other specific interests in fundamental and applied nanoscience. Similarly, the 4th Tri-National Workshop on Standards for Nanotechnology [5] brought together international participation with the goal of harmonizing pre-regulatory measurement standards with increasing focus in the application area of toxicology and identification of measurands. Meetings based on fundamental science and applications such as the International Conference on the Science and Application of Nanotubes [6] also recognize the necessity of good metrological practice in advancing science. The main drivers for investment in development are not necessarily commercial interests, but also policy and regulatory interests. International cooperation as part of OECD and ISO projects stimulates further cooperation and participation, and the outputs of documents and data are increasing momentum in focused areas of nanoscience and nanotechnology. Nano is one of the first real new areas in metrology's long history, and metrologists are learning how to work together in seemingly unconventional groupings and with other technical communities. Looking forwards, metrology and application will be working increasingly closer together.

4.3 Evaluation of Measurement Uncertainty

The quality of a measurement result is provided by the statement of uncertainty. The Guide to the Expression of Uncertainty in Measurement (GUM) [7, 8] provides guidance for evaluating measurement uncertainty so that measurements and uncertainty statements can be compared with each other. Chemical metrology and physical metrology differ in the methods by which traceability to a reference is established, and detailed instructions specific to chemical measurements are outlined in Guide 34 [9]. These Guides on evaluation of uncertainty provide discussion and some specific illustrative examples; the goal is implementation of a common method so that interpretation of the quality of a measurement results is harmonized. Measurement uncertainty is defined as a parameter characterizing the dispersion of the quantity values being attributed to a measurand, based on the information used about the measurement influences. Metrological traceability requires an established calibration hierarchy. Often times the smallest values of uncertainty are associated with measurements that are closest to the direct realization of the unit in the traceability chain. Measurement uncertainty always increases downstream along the chain of measurements. As further comparative measurements are made down the chain, the measurement uncertainty becomes larger due to the introduction of experimental uncertainty components.

The Guides also elucidate some general conventions, such as the practice of reporting expanded measurement uncertainty with a coverage factor of $k=2$, which means that there is a 95% probability that the true value of the measurand lies within the stated range. Other coverage factors can and are used in practice, depending on the application and the intended end-use of the uncertainty value.

The Guides are prepared in a general form in order to be applicable to all measurement capabilities from R&D measurements for primary standards to shop floor routine measurements. Their wide-spread implementation has spurred other more targeted guidance documents specific to calibration laboratories; for example, the European Accreditation (EA) offers instructive documentation and many worked examples relevant to calibration labs [10]. A useful compendium of guidance documents and specific examples for all measurement domains can be located on the BIPM website [11].

Much as measurements of nanomaterials and nano-objects are in stages of development, similarly models of uncertainty evaluation are in early stages of development and uncertainty evaluations for nanoscale measurements are not very common in the literature; targeted implementation of the general Guides is still in development. Many users – both fundamental R&D and applied – would benefit from guidance on how uncertainty evaluations are used in the traceability chain. The effort expended in doing a proper measurement uncertainty budget can be utilized to improve a measurement; namely, by identifying and reducing the largest contributors to the overall uncertainty. In turn, these practices can be used to improve quality and consistency of products and processes. The practice of uncertainty application for nanoscale measurements will eventually become more familiar and similarly develop in niche areas of application such as characterization of carbon nanotubes or other industrially relevant nanomaterials as more labs contribute to the understanding and implementation of measurements.

Measurement results and uncertainty statements are both validated through *comparison* exercises. Several labs perform a prescribed measurement on the same artefact or sample in turn, and the results are compared with each other. The international metrology community has prepared guidelines for CIPM key comparison pilots and participants [12], some of which are also practical for industry round-robins. Comparison validation applies particularly to those measurements having very low uncertainty and/or which are absolute measurements realizing the SI unit, although comparison testing is valuable to evaluate the state-of-play of industrial-level techniques. Comparison data can be analysed by simple visual observation of the plotted data to verify that all labs obtain the same result within the boundaries of their reported measurement uncertainties as expected. In the case of data that is suspected of being discrepant, statistical methods and tools [13–16] are available for more thorough analysis. Identifying the technical reasons behind discrepant or outlier data can make important contributions to our understanding of measurement methods and the sample.

4.4 Metrology & Industry: An Example from Length Calibration

Metrology forms the bridge from science to technology. Calibration and performance evaluation of instruments is a very high priority to manufacturers so that they can produce high quality products and trade their high-tech components on a global scale. The ability to innovate and commercialize new products in nanotechnology depends on the development of measuring tools to provide scientific measurement results and accurate characterization. In some applications, new tools need to be created as existing methods cannot meet surfacing demands. ISO/TC229 Working Group (WG) 4 is tasked with developing specifications and guidance documents for industry characterization of nanomaterials. Some measurement methods being considered by WG4 for particle size characterization are outlined below.

Scanning probe microscope (SPM) instrumentation is applied in almost all areas of nanotechnology and therefore strengthening our understanding of SPM measurement and calibration directly contributes to a very wide technical area. *Validation* of SPM measurements and techniques demonstrates the ability to measure accurately, which benefits industry because it means that in turn, the broad spectrum of all specialized devices and items characterized with these instruments can be considered calibrated and acceptable for the global marketplace. Both SPM and scanning electron microscopy (SEM) instrumentation are typically calibrated via grating artifacts where the measurand is the pitch, defined as the spacing between adjacent lines. Measurement comparisons provide the data and technical consultations with which to improve measurement methods and thereby increase the quality of goods. Comparison of grating pitch measurements made by industry labs, and including measurement results of the national metrology institutes provides those industry labs with an indication of the quality of their measurements, and moreover direct authoritative evidence that their measurements are in agreement with the SI definitions. Some industry round-robin grating-pitch and particle size studies [17, 18] have already been undertaken in addition to the international comparisons amongst NMIs on grating pitch calibration [19, 20]. The observation of some discrepant results demonstrates the need to develop both artifact and documentary standards and measurement protocols to improve the comparability of measurements.

4.5 Key Elements of Metrology Currently in Use

Traceability to the SI for most measurements of nanomaterials is difficult to establish. At this time, method-defined measurements will most likely be the source of reproducible results as research continues to uncover and develop SI-traceable techniques that can be broadly applied. A recent editorial in toxicological literature raises concern for the potential of misunderstanding particle size specifications and the need for clearer definitions [21]. Physico-chemical characterization is a likely area where traceability could be established and appears to be a reasonable starting point for standard protocols and methods. An example of this is measurement of size. Many techniques are available to measure particle size including scanning probe microscope (SPM), scanning electron microscope (SEM), transmission electron microscope (TEM), differential mobility analysis (DMA) and methods employing light scattering techniques such as dynamic light scattering (DLS) and x-ray diffraction (XRD). Direct traceability to the SI-unit of the metre is currently a challenge for many of these measurements because metrological instrumentation is not widely available and the techniques are under development. For a particle diameter measurement, the importance of a well-defined measurand (defined boundary conditions for precise measurements) is demonstrated in the case of the gold reference material produced by NIST where differences in the measured size of the same particle sample are observed depending on the measurement method. Current experience provides explanation for some of the discrepancies, but some differences remain unexplained and are the topic of some interesting on-going research projects. In the interim, metrological microscope instrumentation is under development at many NMIs and techniques are being adapted from analogous macroscale dimensional metrology and particle size characterization.

An important aspect in selection of a measurement or characterization method is the intended end use. If the nanomaterial will be used in water-based solution, then a measurement of the dry sample may have very limited value since it is the size of the particle in solution that will interact with the system under study. Further measurements are required along the process and at the end-point of the study in order to monitor changes. The style of the measurement used to characterize the nanomaterial or nano-object needs to be consistent with the intended study. The SI-traceability chain for general particle-size measurement is not clear at the moment and is mostly dependent on the methods used in practice. Whichever measurement is used, it is important to state the method, and associated detailed information such as calibration, measurement uncertainty estimations and other relevant observations.

In order to rapidly advance our knowledge about nanoscience and technology, it is particularly important for characterization and international standards labs to be able to collaborate and pool results in order to establish standardized characterization protocols and cascades, whereby the relative behaviours of nanomaterials can be determined. The development of a scale of reactivity requires reliable measurement results and reference materials that can serve as reference points along scales of behaviours (toxicity, or other desired characteristics). Examples of materials with positive and negative behaviours are required to establish a scale. A readily available databank would expedite international development of a scale or system in support of predictive models of behaviour and could in turn be used by regulatory agencies. A collective approach allows labs to leverage a knowledge base of material sciences information, characterize the material against a panel of standardized assays, and in so doing facilitate the development and translation of a nanomaterial to application. Indeed one of the challenges facing nanotechnology is the recognized gap between the development of relevant metrological reference standards and identification of which measurands are required for toxicological studies. Similarly, establishing detailed protocols for standard biological assays, and specification of biological media and boundary conditions for use would be a step forward *vis a vis* widely comparable results from toxicology studies.

Setting up this infrastructure of reference materials, calibration and testing labs and providing easily accessible data with which labs can compare hinges on reliable measurements and international cooperation. The NMIs have a culture of cooperation in traditional areas of metrology and their involvement with the documentary standards organisations can expedite formation of tools and mechanisms to support nano development. For example, one of the first steps is to establish infrastructure of recognized measurement capability.

Continuing with the example of size and shape, size measurement is typically referenced to the length unit, the metre; however, measurement of dimensions on the nanoscale with SI-traceability is not straightforward [22]. Instruments with known metrological integrity are sophisticated and not widely available. Conventional methods and definitions for routine measurements and commercial applications are still in development. For example, a particle size measurement made by DLS is challenging to relate in a direct chain of inference to the SI-definition of the metre. Reference standards must be used to calibrate the instrumentation, and the errors such as those associated with the difference in shape from the actual "potato-shaped" particle compared to the perfectly spherical reference need to be taken into account. R&D continues in this area of study, and in the mean time, method-based standards will likely be used to fill the gap until SI traceability can be established. It remains to be seen how some measurements will be made traceable to the SI. For example optical scattering techniques can be used to measure surface features with sub-nanometre resolution. At this point in time, these methods are being considered by both *Comité consultatif pour la quantité de matière – métrologie en chimie* (CCQM) and the *Comité consultatif des longueurs* (CCL). Reference materials and methods are being developed by the length measurement community for microscope-scale calibration. It is important to harmonize the realization of the SI, through optical methods and mechanical methods, so that biases are understood and results can be compared with each other.

Educating and instilling a metrological culture of measuring materials before using them, checking results for unexpected changes, comparing with standards and evaluation of measurement uncertainty would benefit collaborative R&D from the perspective that useful data could be collected from broad sources.

4.6 Redundancy vs. Duplication

Reference materials provide a means to establish a baseline result, or an anchor to compare other materials that have been measured in the same manner. The protocols describing detailed sample preparation and measurement methods [23–25] are important contributors to reliable comparison of measurements made on the same instrument or those made independently by another lab. Establishing the content of protocols is based on measurement repeatability within labs and reproduction of results amongst peer labs. At the formative stage of knowledge, independent demonstration of the same result is not simply redundant; rather it validates understanding of measurement and uncertainty models and provides confidence in the robustness of a protocol and method. Ability to compare measurements made in different labs is essential to moving science and technology forwards; so much so that statistical methods specifically targeted for comparing measurements have been developed by the metrology community [13]. A given measurement becomes "standard" or "conventional" when more and more labs establish capability for the method or technique and they can all demonstrate equivalent results. Establishing conventional methods which are widely known to be reliable is an important step towards the implementation of nanotechnologies. Good metrological practice is essential at every step to keep these developments moving forwards.

4.7 Current State-of-Play and Trends

Metrology for nanoscale applications is in early stages of development. Many measurements are not directly traceable to the SI and GUM-compliant measurement uncertainty statements are rare to find in the literature. In length metrology, metrological microscopes and measurement techniques offering direct traceability to the SI metre [26] exist in a few laboratories – mostly NMIs, and dissemination of SI-traceability to users is often challenging. Standard methods and calibration artefacts are in development and these topics are the focus of much length metrology R&D. Work at the NMIs focuses on diffractometers, SPMs and critical dimension SEM (CD-SEM), and the ISO Technical Committees are attending to documentary standards describing methods, definitions, terminology and quality specifications for artefacts [27]. A contributor to the challenge of disseminating SI-traceability is that many instrument users do not believe they require traceability. Instruments perform well, and so users often neglect traceable calibration exercises and measurement validation. The requirement for traceability becomes more obvious when, for example, a fabrication laboratory needs to change a tool and, in doing so, realizes that newly fabricated parts do not fit in with the rest of the process because the new tool is not measuring with the same reference.

General trends in nanoscale science are still in the process of being uncovered, and it is difficult to identify one single subset of measurands that will be capable of describing any arbitrary nanomaterial. General themes surfacing in nanoscale metrology include: standardization of microscopy tools, mechanical properties of materials, surface interaction/chemistry, thin films, size of particles in air and in water, and biological interactions of nanomaterials. Understanding toxicity for human health and the environment is increasingly a key driver for R&D funding for metrology and instrument development [3] and is a focal point for regulatory agencies and international policymakers [28].

As relates to toxicological R&D, at this point in time the measurement quantities, or measurands, reported in literature and funding proposals describing nanomaterial characteristics has been ad hoc. This poses a challenge for the science and technology community because incomplete or unreliable characterization of nanomaterials results in toxicological data that is inconsistent or impossible to interpret. The community of nanotechnology stakeholders is continually being tapped for input and to share experiences to expedite development of measurement capability to match the needs of toxicologists, and vice versa. Recent meetings and workshops have included industry, academia, and government scientists to discuss minimum requirements of material characterization elements. At the same time, guidance documents on measurement and characterization are in preparation by NMIs and the working groups of ISO/TC229. The ISO/TC229 WG3 documents promote identification of a basic set of measurands (size, surface charge, particle size, aggregation/agglomeration state, composition). It is estimated that many, if not all, will be used for nanomaterial characterization as the field of toxicology for nanotechnology develops. The concept and practical application of a basis set of measurands is in discussion –listing measurands and best implementation of them in a system. A website "The Parameters List" [29] provides up-to-date information and current events in order to maintain open communication on this important topic. One of the goals of the Metrology Study Group of ISO/TC229 & IEC/TC113 JWG2 is to promote and create mechanisms whereby the metrological content of ISO documentation can be improved in support of reliable measurements for nanoscience and nanotechnology. A metrological check-list intended for document review, and a more detailed document providing guidance on what is adequate metrological content of standards are in the process of publication for wider availability outside ISO/TC229. Also discussed are the design and implementation of simple, clear guidance documents, and possible modes which would influence the take-up of metrology in other communities.

References

1. BIPM, IEC, IFCC, ISO, UIPAC, IUPAP, OIML: International Vocabulary of Metrology – Basic and General Concepts and Associated Terms (VIM). International Organization for Standardization, Geneva, Switzerland (2008). JCGM 200:2008 or ISO/IEC Guide 99-12:2007
2. CIPM mutual recognition arrangement (MRA). http://www.bipm.org/en/cipm-mra/
3. National Nanotechnology Initiative (USA). http://www.nano.gov/
4. BIPM workshop on metrology at the nanoscale. http://www.bipm.org/en/events/nanoscale/
5. Tri-national workshop on standards for nanotechnology: measurement & characterization in support of toxicology R&D. http://www.nrc-cnrc.gc.ca/eng/events/inms/2010/02/03/tri-national-workshop.html
6. 11th International conference on the science and application of nanotubes. http://nt10.org/
7. BIPM, IEC, IFCC, ISO, UIPAC, IUPAP, OIML: Guide to the Expression of Uncertainty in Measurement. International Organization for Standardization, Geneva, Switzerland (1995). ISBN 92-67-10188-9
8. BIPM, IEC, IFCC, ILAC, ISO, IUPAC, IUPAP and OIML: JCGM 101:2008 Evaluation of measurement data – Supplement 1 to the Guide to the Expression of Uncertainty in Measurement – Propagation of distributions suing a Monte Carlo method.
9. International Organization for Standardization: ISO Guide 34:2000(E) General requirements for the competence of reference material producers. ISO, Geneva (2000)
10. http://www.european-accreditation.org/content/publications/pub.htm
11. http://www.bipm.org/en/publications/guides/wg1_bibliography.html
12. http://www.bipm.org/en/cipm-mra/guidelines_kcs/
13. Steele, A.G., Douglas, R.J.: Extending chi-squared statistics for key comparison in metrology. J. Comput. Appl. Math. **192**, 51–58 (2006)
14. Cox, M.G.: The evaluation of key comparison data. Metrologia **39**, 589–595 (2002)
15. Wood, B.M., Steele, A.G., Douglas, R.J.: http://inms.web-p.cisti.nrc.ca/qde/downloads/index.html
16. Decker, J.E., Lewis, A.J., Cox, M.G., Steele, A.G., Douglas, R.J.: A recommended method for evaluation for international comparison results. Measurement **43**, 852–856 (2010)
17. Decker, J.E., Pekelsky, J.R., Eves, B.J., Goodchild, D., Kim, N., Bogdanov, A., Wingar, S., Gibb, K., Pan, S.P., Yao, B.C.: Comparison testing of pitch measurement capability on nano-scale length calibration artefacts in industry. Conference on Nanoprint & Nanoimprint Technology, San Jose, CA, 2009
18. Wang, C.Y., Fu, W.E., Lin, H.L., Peng, G.S.: Preliminary study on nanoparticle sizes under the APEC technology cooperative framework. Meas. Sci. Technol. **18**, 487–495 (2007)
19. Decker, J.E., Buhr, E., Diener, A., Eves, B.J., Kueng, A., Meli, F., Pekelsky, J.R., Pan, S.P., Yao, B.C.: Report on an international comparison of one-dimensional (1D) grating pitch. Metrologia **46**, 04001 (2009)
20. Meli, F.: Nano4: 1D Gratings Final Report, CCL-S1 (2003). http://kcdb.bipm.org
21. Maynard, A.: Editorial, Nanoparticles – one word: A multiplicity of different hazards. Nanotoxicology **3**(4), 263–264 (2009)
22. Ehara, K., Sakurai, H.: Metrology of airborne and liquid-borne nanoparticles: current status and future needs. Metrologia **47**, S83 (2010)

23. Decker, J.E., Hight Walker, A.R., Bosnick, K., Clifford, C.A., Dai, L., Fagan, J., Hooker, S., Jakubek, Z.J., Kingston, C., Makar, J., Mansfield, E., Postek, M.T., Simard, B., Sturgeon, R., Wise, S., Vladar, A.E., Yang, L., Zeisler, R.: Sample preparation protocols for realization of reproducible characterization of single-wall carbon nanotubes. Metrologia **46**(6), 682–692 (2009)
24. NIST Nanotechnology. http://www.nist.gov/public_affairs/nanotech.htm; Best practice guides. http://www.nist.gov/public_affairs/practiceguides/practiceguides.htm
25. National Cancer Institute Assay Cascade. http://ncl.cancer.gov/working_assay-cascade.asp
26. Picotto, G.B., Koenders, L., Wilkening, G.: Proceedings of nanoscale metrology workshop. Meas. Sci. Technol. **20**(8) (2009)
27. Ichimura, S., Nonaka, H.: Current standardization activities of measurement and characterization for industrial applications. In: Murashov, V., Howard, J. (eds.) Nanotechnology Standards. Springer, New York (2010). Sect.4.2
28. Organisation for Economic Co-operation and Development (OECD) Working Party on Nanotechnology. http://www.oecd.org/document/8/0,3343,en_21571361_41212117_41226376_1_1_1_1,00.html
29. The parameters list: recommended minimum physical and chemical parameters for characterizing nanomaterials on toxicology studies. http://characterizationmatters.org/parameters/

Chapter 5
Performance Standards

Werner Bergholz and Norbert Fabricius

5.1 Support of Successful Industrialization of Nanotechnology by Anticipative Standardization of Performance Testing: The General Framework

5.1.1 Why Are Performance Standards Needed?

Nanotechnology is emerging now as a technology from the fundamental research stage, so an obvious question to ask is: Is nanotechnology too premature for standardization? The answer will be a clear NO!

The first obvious item for standardization of a technology at the threshold to industrialization is **Terminology and Nomenclature** (Chap. 2). It is a common and unavoidable characteristic of research that there is a relative freedom to define new scientific and technical terms. This may be no problem in the context of science. However when it comes to industrialization, where supply chains and quality management systems need to be set up, a lack of unambiguous terms and definitions is counterproductive, and can be even dangerous. Additionally society expects a responsible and sustainable use of new technologies by controlling potential risks to human health and the environment Therefore governments and national and international bodies are forced to start regulation activities. This requires terminology to describe the items under regulation in a clear and scientific correct way.

In similar manner, new technologies often require new techniques for **Measurement and Characterization** (Chaps. 3, 4 and 6). As with terms and definitions, experience in high tech industries shows that standardized characterization methods are a key component in managing the production of innovative technologies. This includes the measurement of basic material properties as well as the related preparation steps and the presentation of the test results.

W. Bergholz (✉)
Jacobs University, Bremen, Germany
e-mail: w.bergholz@jacobs-university.de

V. Murashov and J. Howard (eds.), *Nanotechnology Standards*,
Nanostructure Science and Technology, DOI 10.1007/978-1-4419-7853-0_5,

Due to the broad public discussion regarding **environmental, health and safety** aspects (EHS) of nanotechnology, these items need to be among the first fields of activity. As stated by the European Commission in its Mandate 409, standardization is one of the building blocks of the "safe, integrated and responsible" approach to nanotechnology. Work on toxicology and screening is mainly within the scope of the OECD (Organisation for Economic Co-operation and Development). Work on risk assessment for chemicals is done by the authorities involved in the implementation of REACH (**R**egistration, **E**valuation, **A**uthorisation and Restriction of **Ch**emical substances), in cooperation with ECHA (European Chemical Agency). In this context the standardization bodies play a key role to provide standardized tools to monitor exposure to the new materials / devices, and how to assess the potential negative effects. These are:

- Methodologies for nanomaterials characterization in the manufactured form and before toxicity and eco-toxicity testing
- Sampling and measurement of workplace, consumer and environment exposure to nanomaterials
- Methods to simulate exposures to nanomaterials

It is also noteworthy that the EHS aspects already receive high level attention and standardization in this area is strongly supported by politics (Chaps. 7–9).

All three aspects are being taken care of in:

- Joint Working Group 1 (JWG 1) of ISO/TC 229 and IEC/TC 113 "Terminology and nomenclature"
- Joint Working Group 2 (JWG 2) of ISO/TC 229 and IEC/TC 113 "Measurement and characterization"
- Working Group 3 (WG 3) of ISO/TC 229 "Health, safety and environment"

The natural starting point and basis for all three areas of standardization is *fundamental research* which is supported and performed by the scientific community, and to a much lesser degree by industry.

The actual structure of nanotechnology standardization is completed by two more working groups:

- Working Group 4 (WG 4) of ISO/TC 229 "Material specifications"
- Working Group 3 (WG 3) of IEC/TC 113 "Performance assessment"

According to the actual business plan of IEC/TC 113 the establishment of further Working Groups is considered. These are:

- Working Group 4 (WG 4) of IEC/TC 113: "Product design"
- Working Group 5 (WG 5) of IEC/TC 113: "Reliability and (material-) FMEA"
- Working Group 6 (WG 6) of IEC/TC 113: "(Nano-) subassemblies and devices"

The structure as shown in Fig. 5.1 indicates, on the one hand, the intention of the involved stakeholders (politics, society, consortia, NGOs and industry) to establish a harmonized system of standards wherever it is possible. On the other hand, the need to address the specific requirements in different industries and product areas is also important. EHS issues are more important for large scale nanoparticle

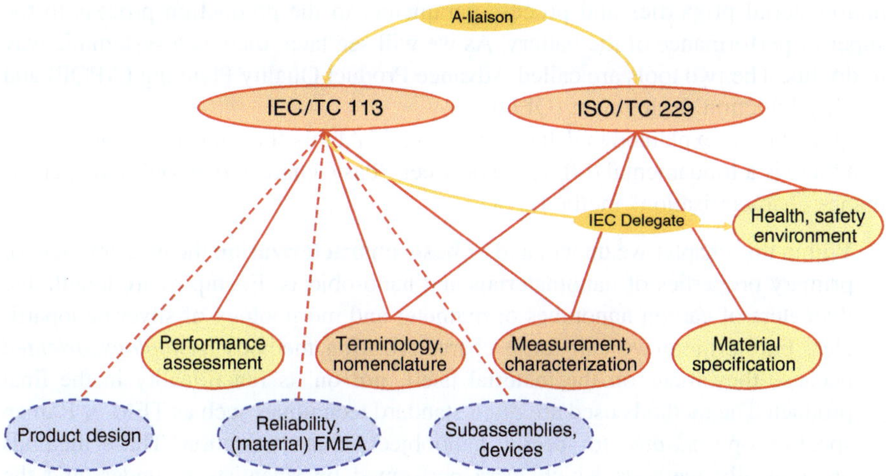

Fig. 5.1 Structure of nanotechnology standardization covered by IEC and ISO with two joint working groups, three working groups (*yellow*) belonging either to IEC/TC 113 or ISO/TC 229 and three future planned IEC/TC 113 working groups (*blue/dashed*). The way the two technical committees co-operate is described in the IEC/ISO directives as an A-liaison. Additionally there is one technical expert of IEC/TC 113 who is also a member of ISO/TC 229/WG 3 "Health, safety and environment"

producers from the chemical industry than for the electronic industry with a well controlled clean room fabrication environment, especially if the nano-objects are part of a device which is hermetically encapsulated.

This chapter focuses primarily on **performance standards** which are in the scope of IEC/TC 113/WG 3 "Performance assessment." They are intended to support the fabrication of new innovative products with extraordinary high performance enabled by the use of nanotechnology. Therefore, they support commercialization of scientific results by providing standardized methods to qualify nanomaterials and control nano-related production processes. Nanotechnology – enabled products are to be developed and produced for a **specific purpose**. In other words, the product must be specified in terms of its *performance from the perspective of the customer/user*, which is a completely different point of view than the *view of the engineer or scientist* for nanotechnology materials or products.

To illustrate this point, we examine the case of a nano-enabled battery. The main relevant performance measures from the users' perspective are energy storage capacity, recharging time and the maximum power that can be drawn from the battery. Additionally the user will be interested in the durability,[1] i.e. the degradation of the mentioned performance indicators with the number of charge/discharge cycles. To design and produce such a battery, there has to be a model which relates

[1]Durability (as defined in IEC 60896-21 Ed. 1.0): ability of an item (battery) to perform a required function under given conditions of use and maintenance, until a limiting state is reached.

Note: A limiting state of an item (battery) may be characterized by the end of the useful life, unsuitability for any economic or technological reasons or other relevant factors.

Definition according to IEC 60896-21 Ed. 1.0 (2004).

nanomaterial properties and process parameters in the production process to the superior performance of the battery. As we will see later, there is a systematic way to do this. The two tools are called Advance Product Quality Planning (APQP) and Quality Function deployment (QFD).

QFD will be explained in detail in Sect. 5.2.3. At this stage it is important to note that there is a fundamental difference between "basic characterization" and "performance characterisation" methods:

- Within this chapter we understand as **basic characterization** the measurement of primary properties of nanomaterials and nano-objects. Examples are length and diameters of carbon nanotubes or diameter and morphology of silver nanoparticles. Furthermore we call these characterization methods *technology oriented* because they focus on the material itself, **not** on its functionality in the final product. The methods used are often standard techniques such as TEM or Raman spectroscopy *adapted* to specific nanoobjects characterization. These methods are typically methods which were performed by scientists to understand the nature of nanomaterials and nano-objects. In this respect these methods are the basis for technical break-through and the creation of innovations. Very often, these methods are not suitable for fabrication control and quality assessment.
- The requirements for manufacturing are completely different. In a fabrication line we can assume that the materials used are more or less understood and are delivered in an acceptable quality in accordance to a material specification. What we need additionally is a fast and easy-to-perform method to control the quality of these materials in the fabrication line, with respect to the **specific nano-enabled final product property** that the nanoobjects are intended for. These methods need to be sensitive for variations of material properties which directly influence the product performance. Therefore, we call these tests **product-oriented performance tests**. It is often the case that performance characterization methods are very specific to the *application* [1] for which the nano-enabled product is used. To illustrate this point, we give a number of specific examples for such performance characterization methods. Assume that the nano-enabled product is a battery for which a nanomaterial added to one of the electrodes enhances the energy density of the battery.

  – A product related test method for the subassembly "anode" could be a standardized simple and fast procedure to produce test anode inserted into a standard battery test assembly, which is then characterized for energy storage capacity.
  – Assuming further that the nano-enabling property is the surface to area density of the nanomaterials employed. A "surrogate" test which indirectly relates to the storage capacity, but is directly sensitive to the surface to volume ratio, could be used, such as (possibly) the current to voltage characteristics of the electrode in an electrolyte under standard test conditions, or in fact any other property which depends on the absolute surface area.
  – Another embodiment of such a product-related test for the given example is any test which is sensitive the surface-to-volume ratio of the nanomaterial, at even an earlier stage when the nano-material is as-received from the vendor. At that stage, one can envisage a test in such a form that a substance which is

adsorbed at the surface of the nanomaterials is added to the solvent in which the nanomaterial is dissolved, and the rate of adsorption is monitored by a suitable quantity, such as conductivity, color or ph-value of the solution.

- It is common to all these performance tests that they are quick and easy to perform and that they yield a control parameter which is a predictor for the desired product property. Such a parameter is called a **key control characteristi**c (KCC) with respect to the final product. The correlation between the KCCs at different stages of the fabrication and the desired final product property has, of course, to be established first.

One of the most important "truths" of industrial production is, that the quality, i.e., the performance of a product according to specification, cannot be "tested into a product." Rather, that the manufacturing process has to be managed to *ensure* that the product has the desired performance. This is the purpose of the *indirectly related performance test parameters* or **K**ey **C**ontrol **C**haracteristics[2] (KCCs) (see below).

Such indirect performance standards for electronics and electric nanotechnology enabled products, i.e. manufacturing and material related performance parameter are the scope of **IEC/TC 113/WG 3** (see Fig. 5.2).

There is a need for standards relating to every stage of the value adding chain. WG3 is at present covering Material Parameters and Production Process Parameters, which have a significant impact on the product performance parameters. In the near future, other stages will have to be addressed.

The nano-enabled sub-assemblies, such as a PV cell made from nano-objects, and nano-enabled product, such as PV module assembled from nano-enabled PV cells require standards too. The development of such standards is planned in IEC TC 113 **WG5**.

In addition to performance, two additional essential product characteristics are *reliability and durability*. The planned working group IEC TC 113 **WG4** will deal with reliability and durability standards, the logical connection to the performance standards in manufacturing is obvious. It only makes sense to do reliability and durability testing if the manufacturing process is *well-defined* (i.e. standardized, at least within the company!), with respect to all parameters which affect the final product, and that it is stable.

Another principle of high quality manufacturing is that good quality starts with a *robust design* both for the product and the production process, therefore a **WG6** is planned to complement the "Performance Management" along the whole life-cycle of a nano-enabled electronic product by creating a standard for principles/requirements of/for the design process, along the lines of the Quality Management Standards ISO 9001, Clause 7 [2].

---

[2]Key control characteristic (term will be defined in IEC/TS 80004-9) process parameters for which variation must be controlled around a target value to ensure that a significant characteristic is maintained at its target value.

Note: KCCs require ongoing monitoring per an approved Control Plan and should be considered as candidates for process improvement.

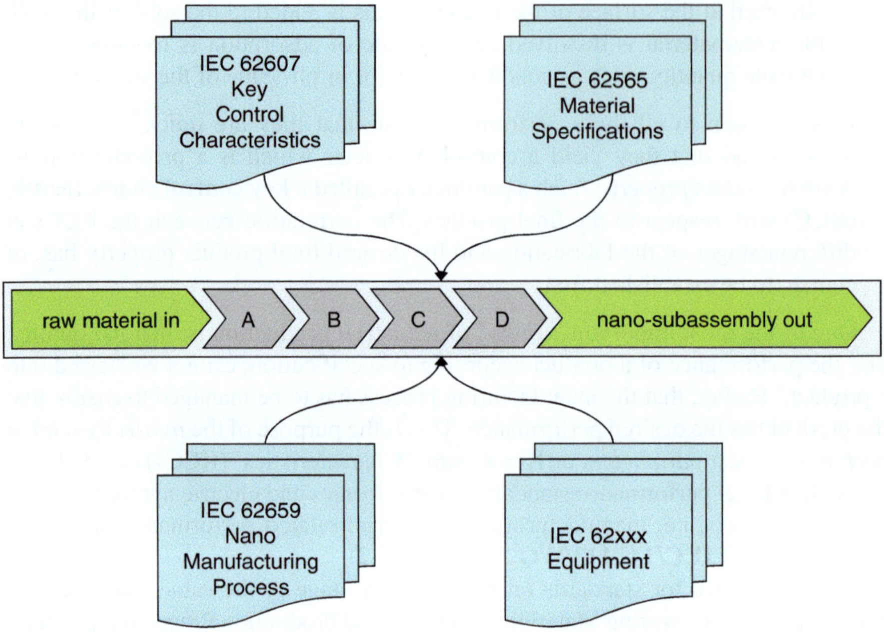

Fig. 5.2 High quality nano-manufacturing requires the simultaneous use of four groups of standards: Material specifications, key control characteristics, equipment and processes. Because the product performance for nano-enabled products is strongly dominated by the use of nanomaterials or more generally by nano-objects material specifications and the accompanied key control characteristics plays an exceptional key role in this scheme. For the actual projects see Table 5.1

5.1.2 Anticipative Standardization

The different types of standards along the lifecycle have been explained, what remains to be addressed is to make the point that anticipative standardization is a good idea. Industrial production of nano-enabled products is at a very early stage, while the predicted volumes and ubiquitousness of nano-enabled products are significant.

To avoid confusion about "not in production," we emphasize that "normal" microelectronic products, manufactured by standard planar lithography with minimum feature sizes <100 nm have been in production since about 2005, but they are *not* within the scope of this paper. For those products hundreds of standards exist already, both with IEC and SEMI, which has about 700 consortium standards in its portfolio, most of which are being intensively used by the semiconductor industry. The scope of this paper will be those nanotechnology products which are made by bottom up technology (i.e. start with nano objects and assemble them) rather than top down technology (i.e. create a macroscopic layer and then structure it by lithography tools).

Although we will not deal with microelectronics, it serves a good example how anticipative standardization can support industrialization of an emerging technology. In the 1960s and the beginning of the 1970s, there was practically no standardization

Table 5.1 Actual projects in IEC/TC 113/WG 3

Nanomanufacturing – Material specificati on	IEC/TR 62565-1: Nanomanufacturing – Material specification Part 1: Basic concepts
	IEC 62565-2-1: Nanomanufacturing – Material specification Part 2-1: Carbon nanotube materials – Blank detail specification for single wall carbon nanotubes
	IEC 62565-2-2: Nanomanufacturing – Material specification Part 2-2: Carbon nanotube materials – Detail specification for single wall carbon nanotubes for application xyz
	IEC 62565: Nanomanufacturing – Material specification Part 3-1: Material 3 – Blank detail specification
Nanomanufacturing – Key control characteristics	*IEC/TR 62607-1: Nanomanufacturing – Key control characteristics Part 1: Basic concepts*
	IEC/TS 62607-2-1: Nanomanufacturing – Key control characteristics Part 2-1: Carbon nanotube materials – Film resistance
	IEC/TS 62607-2-2: Nanomanufacturing – Key control characteristics Part 2-2: Carbon nanotube materials – KCC 2-2
	IEC/TS 62607-3-1: Nanomanufacturing – Key control characteristics Part 3-1: Luminescent Nanoparticles – Quantum efficiency test
	IEC/TS 62607-3-2: Nanomanufacturing – Key control characteristics Part 3-2: Luminescent nanoparticles – KCC 3-2
Nanomanufacturing – Process specification	IEC 62659: Large scale electronic manufacturing [10]
	IEC 62624: Test methods for measurement of electrical properties of carbon nanotubes
Nanomanufacturing – Equipment specifications	*Projects to be established in the future*

The grey areas in the table shows the existing projects, the white areas are planned projects for the near future (italics letters)

in the emerging microelectronics industry. As a consequence, problems started to accumulate as the production volumes expanded, e.g., material shortages, inconsistencies between materials and processes, unnecessary high transaction costs, and quality problems [4]. Over time, lessons were learned, and when the microelectronics industry moved from 200 to 300 mm diameter wafers in the second of the 1990s, essential steps in the value chain were standardized *before* pilot production or even development started. It is estimated that this anticipative standardization has saved many billions of dollars. Currently, photovoltaics is going through the same "painful" learning process, after the initial industrialization period (with annual growth rates of 45% over 10 years) happened without any appreciable standardization. The industry is now cooperating to create a suite of standards which will support the next wave of mass production in factories with gigawatt annual output.

Therefore it appear more than reasonable to apply this "recipe for success" to the emerging industry for nano-enabled electronic products. A widespread stereotype about standards is that they slow down innovation. Microelectronics is a highly innovative industry, so there is overwhelming evidence that standardization supports rather than hinders innovation.

In addition to the need to cover all stages of the value chain to manage the performance of the end products from design to sub-assembly, and to the issue that ideally standards exist before large scale production starts, there is a third important aspect that has to be covered in defining the framework for standardization work, namely **how are existing and the new standards aligned**? This topic will be analyzed in Sect. 5.1.3.

5.1.3 Nanotechnology Standards and Existing Standards

In the near future, nanotechnology will be applied to many different electric and electronic product categories, such as displays, energy storage devices (batteries, capacitors), energy production devices (such as PV cells) and others. For most of these, products, technical committees in IEC already exist, and hence in many cases performance standards for the final product already exist too.

To ensure that there is no "reinventing the wheel", it is essential that intensive communication is set up to all relevant TCs in IEC, ISO and other standardization organisation such as IEEE, SEMI, etc., as visualized in Fig. 5.3.

This reflects the fact that nanotechnology is a "**cross sectional or umbrella** enabling" technology which can be potentially applied to a large number of different product categories (the term umbrella technology is used in Chap. 1). To link to all relevant TCs at this point in time would be complex and probably not make much sense. Rather, care has to be taken such that in the practical standardization work those product categories are prioritized for which industrialization is imminent *and* that liaisons are established and maintained to the respective technical committees.

Similar reasoning applies to other cross sectional technologies, such as clean room technology, waste disposal, etc.

5.2 How to Practically Create Standards Through the Whole Value Adding Chain/Supply Chain

5.2.1 Quality and Process Management

In the electro-technical and electronic industry, the level of quality has been increased by a factor of 100 or more, if measured by the number of delivered defective products or the number of early failures (failure rate during the first 1,000 h of operation) [5]. As can be seen in Fig. 5.4, both indicators decreased continuously from a level of a few 100 to a small single digit numbers in the time period from 1985 to 2000.

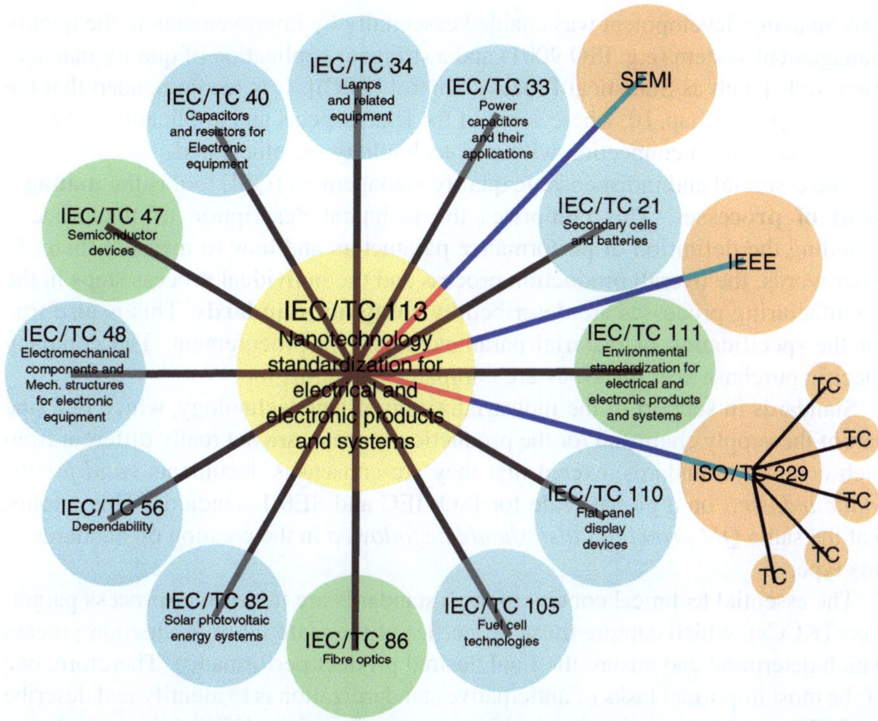

Fig. 5.3 Overview of IEC Technical Committees and other organisations for which nanotechnology is potentially relevant. TCs in IEC which are blue and green (existing formal liaisons). Formal external liaisons are also established to SEMI and IEEE. The very close relation to ISO/TC 229 ensures the communication to ISO/TCs within the Nanotechnology Liaison Coordination Group (NLCG)

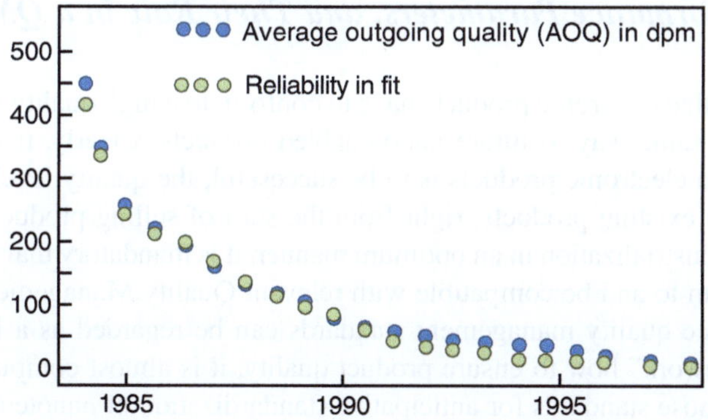

Fig. 5.4 Quality improvement in the Chip Industry, as published by the Siemens Semiconductor Division for their Microelectronic Products [5]. Reliability is quantified by the failure rate measured in FITs (Failures In Time). 1 FIT is one failure per 10^9 device hours. Average Outgoing Quality (AOQ) is measured as one defective part per million delivered to the customer

This amazing development was enabled essentially by improvements in the quality management system (e.g. ISO 9001) and a stringent application of quality management tools [such as Statistical Process Control (SPC)]. It is recommended that the reader refers to Chap. 10, where some of the legal aspects and implications of quality management in connection with nanotechnology are elucidated.

One essential and indispensable quality management (QM) tool is the **management of processes**. This comprises the technical description of the process, including the definition of performance parameters and how to measure them. In other words, the overall production process and the individual process steps in the manufacturing processes are described by **company standards**. This is also true for the specification of material parameters used in procurement. The company specific purchase specifications are company standards, too.

Standards in support of the industrialization of nanotechnology, which describe part of the supply chain and /or the production process, are not really different from such company standards, except that they are consensus documents *valid for the whole industry*, on a global scale for ISO, IEC and SEMI standards. This implies that the same *QM principles also should be followed* in the creation of standards of this type.

The essential technical content of such standards are material or process parameters (KCCs), which capture those properties of the material or production process which determine and ensure the final desired product performance. Therefore, one of the most important tasks of anticipative standardization is to identify and describe the KCCs, and support the practical implementation of the KCCs by standards.

In the following Sect. 5.2.2 we will therefore describe the method for identifying the essential direct and indirect performance parameters in connection with the manufacturing of nano-enabled electronic products.

5.2.2 Key Control Characteristics as Direct and Indirect Performance Parameters, and Their Role in a QM System

We have seen that electronic products have to conform to a high quality standard; this applies in the same way to future nano-enabled products. Clearly, if the industrial launch of nano-electronic products is to be successful, the quality level must be the comparable to existing products, right from the start of selling product. In order to support the industrialization in an optimum manner, it is mandatory that the standards should conform to and be compatible with relevant Quality Management standards. Moreover, since quality management standards can be regarded as a kind of "best practice framework" how to ensure product quality, it is almost compulsory to take into account those standards for anticipative standardization of nanotechnology.

Although the ISO 9001 standard is the most widely used standard for a QM system, for some industries it is not deemed sufficient, since it sets to loose and non-specific requirements either for safety/environmental aspects, or in terms of making sure that specific requirements to ensure the economic viability of an operation are met. Such standards exist e.g. for the Aerospace industry, the medical and food industry, and last but not least for the automotive industry. One of the most important and widely used standards in the electronic industry is the ISO TS16949, which is **the** worldwide QM standard for the automotive industry. TS 16949 contains the complete ISO 9001 QM standard, but has a high number of additional features and QM-tools, which according to the authors experience are very instrumental to promote quality from the drawing board stage to mass production [6].

The majority of electronic product types which are likely to benefit from nano-enabled materials or subassembly will have automotive applications, for which conformance to TS 16949 is mandatory. Therefore, it makes sense to align our realization strategy for nanotechnology performance standards to the stipulations of this particular standard, which will ensure almost "automatically" that companies applying the standards conform to TS 16949.

TS 16949 demands (among many other items) that:

- Customer expectations, i.e. the product performance parameters must be clearly defined.
- The manufacturing process parameters which determine the product performance parameters are identified (e.g. by the QFD method) and that the processes are stable, as defined by Statistical Process Control (SPC) principles. Such parameters are the Key Control Characteristic (KCCs) mentioned earlier. Details can be found in a supplement to TS16949, the Production Part Approval Process (PPAP) document. Often, these parameters are called **special characteristics**.
- The material parameters which determine the product performance parameters are identified and controlled in the same manner.
- Design of products and processes is done in such a manner that the product design makes sure that quality, durability and reliability is already "designed" into the technical concept, construction details and into the production process, which includes the anticipative identification of KCCs and whether manufacturing processes and materials are able to conform to the specified ranges for the KCCs at the various stages of the value adding chain. This process is described in another supplementary document to TS 16949 (APQP). So the APQP process is like a blueprint for how to make sure that the quality of the final product is already engineered into the product at the planning or design stage. In other words, it is a proven "recipe" how to engineer the necessary quality into a nano-enabled product when it is most efficient, namely right from the beginning.

The structure of existing and future working groups in TC 113 reflects this (compare Fig. 5.1).

What is left in the description of the QM-based to strategy for anticipative standardization, is to describe how the product performance – sensitive parameters for the production process and the materials are identified. A well established method to implement this is the QFD method mentioned earlier, which is described in Sect. 5.2.3.

Work on standards which follow this concept, and center around KCCs is already underway, the details will be described in Sect. 5.3.

5.2.3 The Quality Function Deployment Method to Identify KCCs

The QFD method [7] is essentially a matrix method in which the **performance requirements of the users for the final product** are written in the first column of the matrix, and the **technical specification requirements for the nano-subassembly** are written in the first row, as shown in Fig. 5.5a. The matrix elements give information on how strong the influence of each of the technical subassembly parameters is on each of the final product parameters (as indicated by numbers between 0 and 9). The technical parameters with the highest column sums obviously have the largest influence on the performance parameters. In this way, a systematic prioritization of the parameters as to their relevance for final product performance can be made. The determination of the appropriate number for each of the matrix elements is a matter of engineering expertise.

There is no "mechanical" rule how to select the KCC parameters. What has to be noted however, that it should be ensured that also those nano-subassembly parameters are selected which have a comparatively low sum, but which are the **only one(s) to represent one of the product performance parameters**.

The procedure to identify parameters should start with all technical parameters which appear to be relevant (so that no important parameter is overlooked) but in the end the aim should be to identify only a few should be defined as KCCs. As an example for a product we consider a battery which contains nanomaterials in one (or both of the electrodes). The product performance parameters could be the storage capacity, the maximum load current and the number of charge/discharge cycles until the storage capacity degrades by 50%. The subassembly in this specific example is one of the electrodes. The KCCs for the electrode subassembly could be the performance of the electrode in a standard test battery set-up. Once the KCCs for the nano-subassembly are identified the next stage is to identify the KCCs for the manufacturing process, as outlined in Fig. 5.5b. Now the technical subassembly parameters are in the first column, and the KCCs for the manufacturing process are the first row, and the sums of the columns and rows are used to identify the most important process control parameters as the KCCs.

The next step (Fig. 5.5c) is the identification of the nanomaterials KCC parameters, in an analogue fashion. An overview of this cascaded QFD process is given in Fig. 5.5d. If needed, the process can be further cascaded to the manufacturing process at the site of the nanomaterials manufacturer.

At this point in time we are not aware of a concrete example that a cascade of such KCCs has been demonstrated for any nano-enabled electronic product. We therefore use, for illustration purposes, a practical example from microelectronics.

The product in this example is a memory module for a PC. The subassemblies are the eight DRAM (dynamic random access memory) microchips that go into one memory module. The product performance parameter is the early fail rate of the DRAM modules, most of those failures are due to gate oxide breakdown in one of the memory cells of the module. The subassembly KCC is the retention time failure

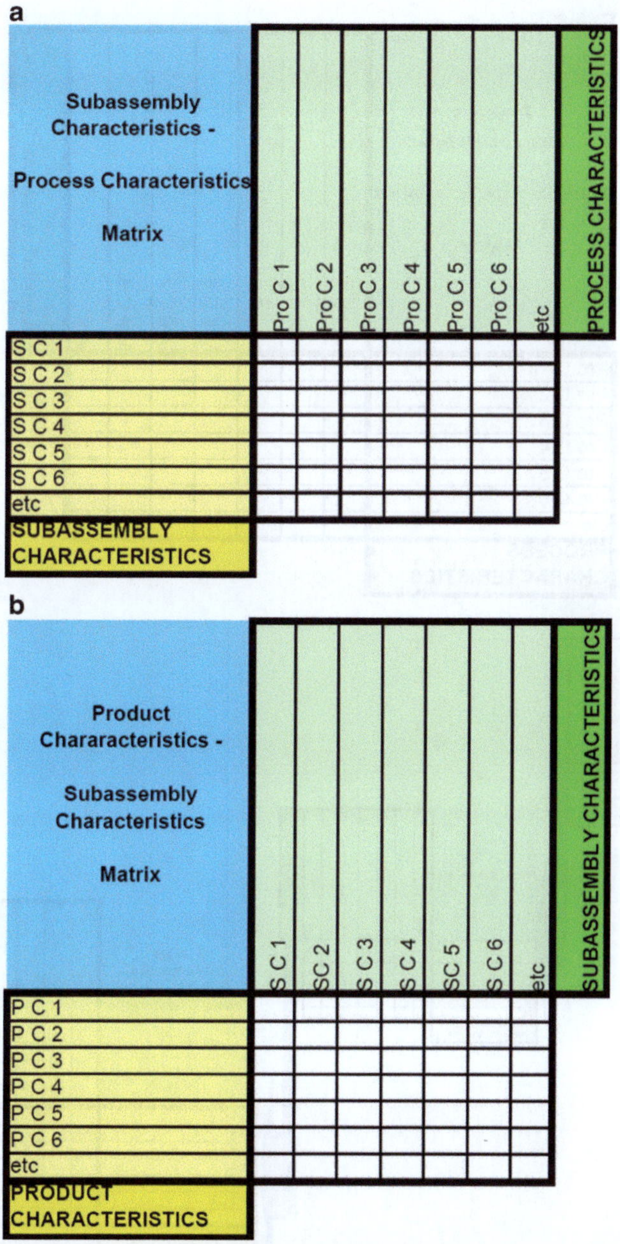

Fig. 5.5 (**a**) QFD Matrix to identify the KCCs for the nano-enabled subassembly, with the nano-enabled product parameters as the input. (**b**) QFD Matrix to identify the KCCs for the subassembly manufacturing process, with the subassembly KCCs as the input

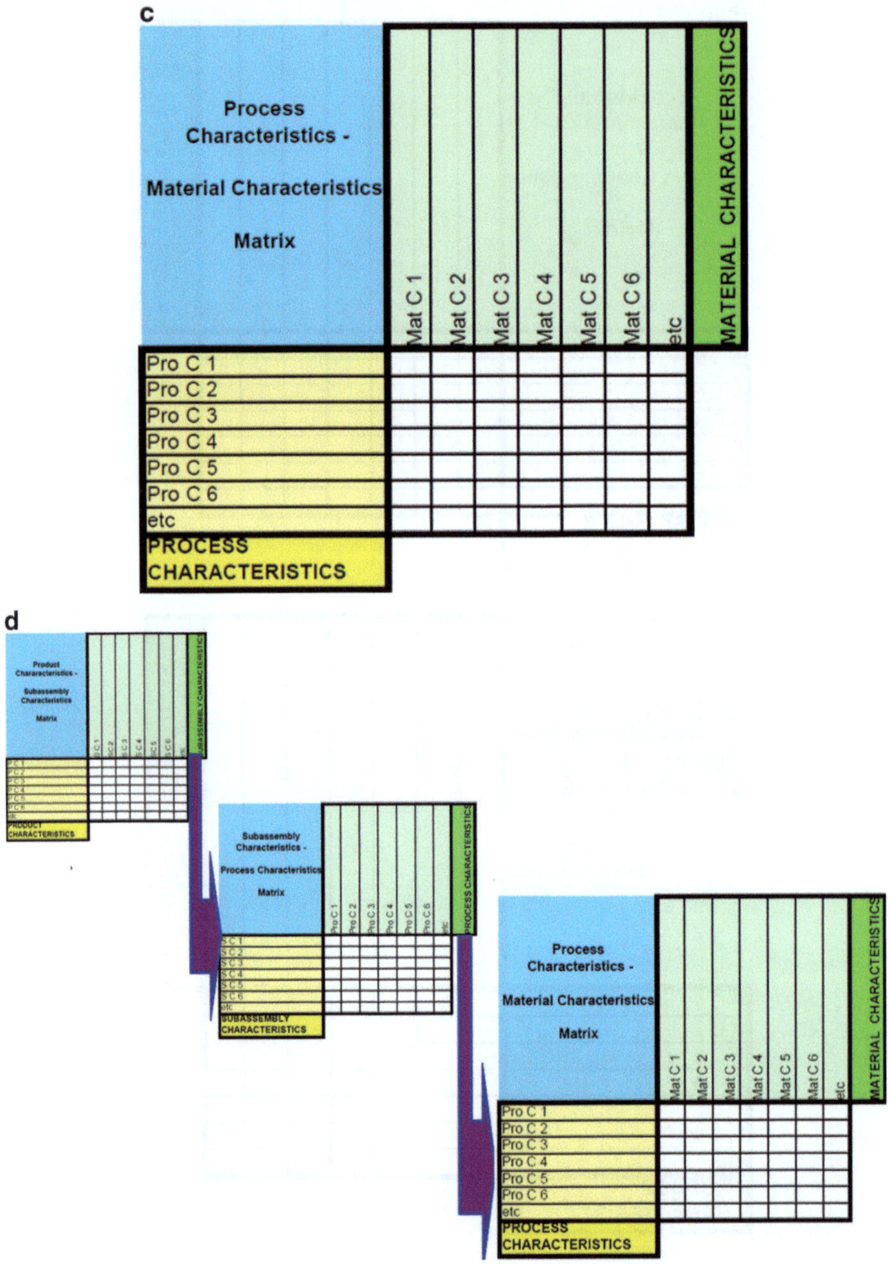

Fig. 5.5 (continued) (**c**) QFD Matrix to identify the KCCs for the nanomaterial, with the nano manufacturing process parameters as the input. (**d**) Overview of the complete QFD Process

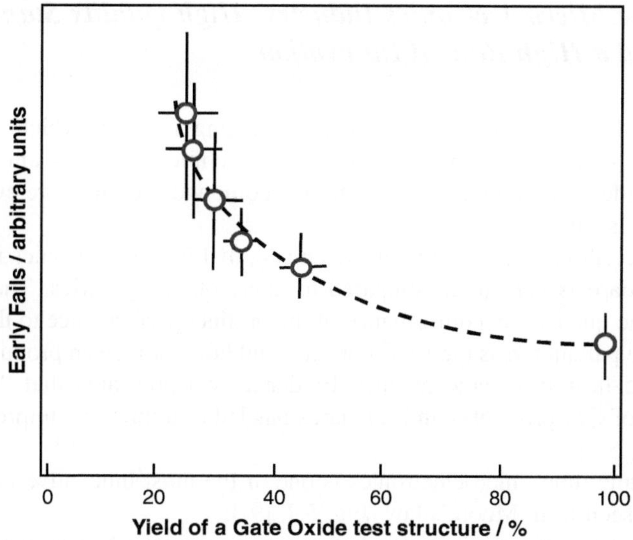

Fig. 5.6 Correlation of the early fail rate of DRAMs with the short gate oxide test for silicon material quality [8]

statistics in the test of the individual DRAM chips on a wafer level. The relevant production process parameters are the metal contamination values of the gate oxidation process and the cleaning efficiency of the cleaning process step before gate oxidation. A third production KCC is the stability of the oxide etch rate before gate oxidation. Finally, one of the relevant material KCCs is the density of voids in the silicon wafer starting material. This quantity was difficult to determine for a long time, in fact before 1995 it was a "hidden" material parameter.

As a "surrogate" test, a short gate oxidation test for raw wafers had been developed, the yield in this quick test correlates with the early fail rate of the product, and thus served a material KCC with high relevance for the reliability of the final product [8] (Fig. 5.6).

As was established later, this KCC can be related to the pulling rate for the silicon crystal from which the wafers are manufactured.

5.3 Nanoelectronics Standardization: First Steps and Practical Experience

In this section, we will detail what has already happened in nanoelectronics standardization, while keeping in mind that the experience collected in nanoelectronics should be usable for other industry segments in which innovative nano-enabled products will play a role. So, in a way, implementing performance standards and a QM based standardization strategy can be regarded as a "pilot project" for the standardization in other areas of nanotechnology.

5.3.1 The Microelectronics Industry: High Quality Standards and a High Rate of Innovation

It has been frequently argued that in an emerging industry which uses a new technology, standardization will slow down technical process and should be actively avoided. Therefore, we return to this point and present more evidence that the contrary is true.

In the preceding section the point was made, that in the electronics industry the quality standard is very high compared to many other industries. The "hub" for managing the quality, i.e. conformance of the product performance to the specified performance parameters is the KCC concept, and how for a given product these can be identified in a systematic manner. Evidence was presented that the stringent application of QM principles and standards has led to impressive improvements of quality (Fig. 5.4).

At the same time, microelectronics is one of the most innovative industries, as can best be seen from Moore's law (Fig. 5.7, [9]).

The decrease in the minimum feature size over almost 40 years would not have been possible without **constant innovation** in terms of materials, processes, and equipment. If anything, standardization has supported the innovation rather than hindered it [4].

Since the situation in nanotechnology is very similar to what it was like in the microelectronics industry 40 years ago, it appears appropriate to select the electronic and electro-technical industry as the industry segment most suitable to "pioneer" nanotechnology standardization.

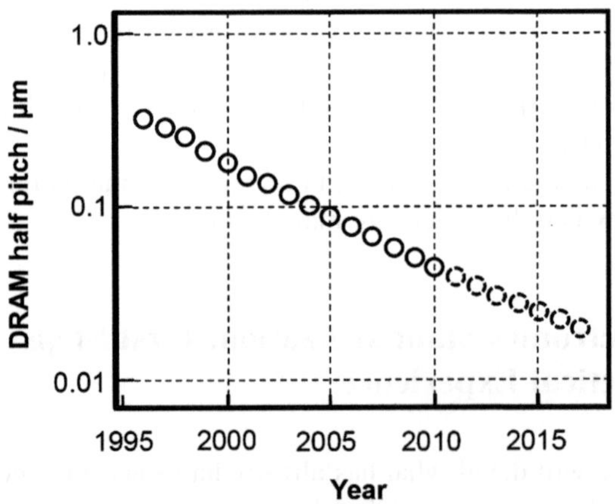

Fig. 5.7 Illustration of Moore's law: By shrinking the minimum feature size over the last 2 decades, the number of transistors (functions) per chip doubled every 18–24 months, while the price for the product has essentially stayed constant or even decreased. The data have been retrieved from the publications of the ITRSl [9]

5.3.2 *Safety Aspects: Cleanroom Technology, Small Quantities of Nanomaterials, Encapsulation of Nanosubassemblies*

Before turning our attention to the concrete details of how the KCC concept is being implemented in IEC TC113 work, it is in order to mention that the electronics industry can contribute also to progress in the EHS area (Chaps. 1, 8–10). This is a significant point, since nanotechnology has a potential negative image and acceptance problem by the public. Although as of now there is no compelling evidence that nano-objects as such are detrimental to human health and the ecosystem. Quite the contrary, many unintentionally made nano materials have been around for decades, examples are carbon black and syton polish for polishing silicon wafers, and there is no evidence for harmful effects. Yet, there is a growing feeling in the public that nanotechnology is not really safe.

The risks that are envisaged, and which are addressed in the projects of ISO TC 229 working group 3, are:

1. Exposure at the workplace when the nano objects are in their dispersed form (occupational risks)
2. Exposure during use of nano-enabled products (consumer risks)
3. Negative ecological effects at the end of the lifetime or during accidental breakage of nano-enabled products (environmental risks)

For all three risk modes, electronic products offer an advantage over other nano-enabled normal product categories:

The *exposure at the work place* is normally not easy to measure since in the normal work environment there is already a large concentration of nano objects in the ambient air, even more so where there is a lot of traffic, where people smoke or where combustion or dust intensive activities take place. Therefore, workplaces have a severe background problem for measuring the exposure to nanomaterials.

Production of electronic material for a large part takes place in a cleanroom, or at least in a controlled environment, where the background level of nanoobjects is well-defined, and therefore very sensitive measurement of exposure is possible. An additional advantage is that cleanroom technology includes well proven methods to put up a **barrier** against small particles. Normally the objective is to protect *the product from the particles mainly shed by humans*, equipment or from other particles sources. The same techniques and the practical experience can be used to protect workers in the work environment from the nanoobjects which might "escape" from the nanomaterials or partially processed nano-subassemblies or products, since the barrier functionality works in both directions.

Exposure of the users (consumers) to nanoobjects from the product is minimal, since most products will only contain minute amounts of nanomaterials, and the nano-enabled devices will normally be sealed from the environment, since the objects will not remain functional unless protected from the environment, in particular from humidity.

End of life emission of nanoobjects into the environment is unlikely, since nowadays as a rule electronic scrap is separated from normal garbage for recycling. The escape during accidental breakage is, in most cases, not critical either since the nano objects will be in an aggregated form, and even if dispersed during the accident, the quantity of material released will, in most cases be negligible.

Therefore, nano-enabled products can also act a **low risk pilot product** category to test in the work place and in the field how to control nano-objects and as test cases whether or not there are undiscovered risks associated with nanotechnology.

5.3.3 The Experience So Far: Current Projects in IEC TC113

With the arguments outlined in the preceding sections in mind, IEC/TC 113 "Nanotechnology standardization for electrical and electronic products and systems" was founded in 2006. The standards deliverables will focus on components or intermediate assemblies that are created from nano-scaled materials and processes for electrical or electro-optical applications.

Potential applications include electronics, optics, magnetics, electromagnetics, electroacoustics, multimedia, telecommunications and energy production and storage. IEC/TC 113 focuses on products which use nano-electrotechnologies in one or more of their subassemblies or during the fabrication process as shown in Fig. 5.8. The committee will produce standards, technical specifications and technical reports to guide manufacturers and customers in situations where it is necessary to

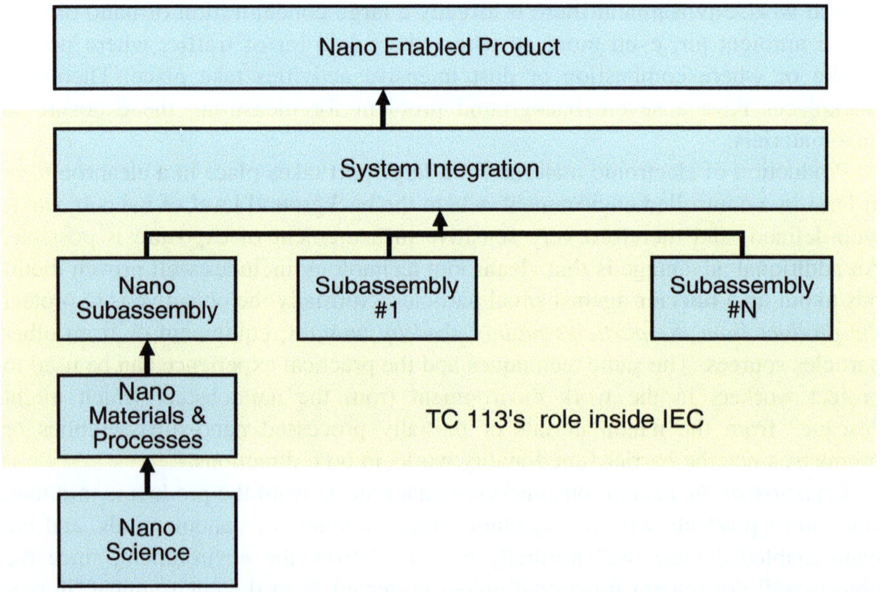

Fig. 5.8 Overview of IEC TC 113 basic role within standardization of electronic and electrotechnical products

use an emerging technology under the absence of complete knowledge. By doing so the products shall be optimized in:

- Life cycle performance
- Reliability
- Operational safety

Standards developed in IEC/TC 113 will address all stages of the economic model:

- Basic (pre-competitive) technical research
- Product development from initial design to prototype manufacturing
- Fabrication including initial deployment and large scale high volume production
- End use (operation) by customer/consumer
- Product end-of-life disposing and recycling

The strategy for standardization was based on the concept of supporting industrialization of nano-enabled electronics by a QM based strategy, as outlined in the previous two sections.

On the basis of common engineering sense, the decision was to start activities not at the end of supply chain, but at the two first stages of the value-adding chain. This work is allocated in IEC/TC 113/WG3 which addresses the most important elements of nanomanufacturing:

- Materials: Today nanostructured materials are the essential part of nanotechnology. Nearly all existing nanotechnology products derive their performance from the use of nanostructured materials. The structure is on the nanoscale either because this structure has been formed:
  - During the material fabrication process
  - During the application of material onto some kind of a substrate (homogenous thin layers/nanostructured layers)
  - By chemical, or physical surface treatment of homogenous material (e.g. removal of material by a atomic microscope tip) or
  - By nano-objects (e.g. particles, nanotubes, nanorods) in solutions, mixtures or deposited as agglomerates or aggregates onto substrates

So far there is no systematic way to specify those materials that can be certified by ISO 9000 as it is required for high quality fabrication.

The IEC/TC 113/WG3 approach to solve this problem is the concept of "blank detail and detail specifications" in conjunction with "key control characteristics" as described below.

- Equipment: The manufacturing and use of nanostructured materials may require special equipment. At this point in time, such equipment is available and already used in the chemical industry, microelectronic industry and microsystem industry. So far there are no projects in IEC/TC 113 regarding equipment. This may change soon. In the long term perspective, it is anticipated that self assembling processes and in the very long term the use of nano-assemblers will initiate standardization activities in this area.

- Processes: Attention is required to processes. Even if the processes to fabricate nanomaterials today are company intellectual property and confidential we assume that this will change in the future if the main added value will come from the design of products and the use of nanomaterials for special applications. The companies will need special standardized processes which are compatible with other well defined processes. The first step in this direction is a IEC/IEEE joint standardization project addressing the use of nanomaterials in large chip facilities. One example of the addressed issues is the compatibility with the classical CMOS process. It is easy to rationalize that it is not acceptable that there is a risk that the nanomaterials contaminate a multi-billion dollar facility, disturb the fabrication flow and reduce the yield of chip fabrication.
- Key Control Characteristics: Material specifications are a difficult matter. Especially if the quality requirements are very high it is probable that small deviations in the material will influence fabrication yield as well as performance and reliability of the final product. Even in the case of the well known and well defined crystalline silicon used in the microelectronic fabrication there are hidden parameters which have an influence to the fabrication, performance and reliability of the final product (compare the example given earlier, Fig. 5.6, for which an important hidden parameter was only discovered after about 20 years of intensive work). Hidden parameters might be much more important for nano-structured materials. The properties of such materials are not only dominated by their chemical composition, but much more due to imperfections of their nanostructure and very small amounts of impurities. The strategy is that these materials shall be specified as detailed as possible or practical. Additionally special properties shall be measured which are sensitive to material changes which influence the intended function of the final product. These properties are "pragmatic" or special product-related "key control characteristics" because they are directly related to applications, as indicated in Fig. 5.9. The intention is that the combination of the material specification and such product-related "key control characteristics" can define the material "well enough" for high quality fabrication.

5.3.3.1 Example 1: IEC 62565-2-1 Nanomanufacturing – Material Specification Part 2-1: Blank Detail Specification for Single Wall Carbon Nanotubes

Activities were focused on the material which is seen by many experts as the most likely candidate for which industrial application will emerge, carbon nanotubes. A first survey (of the technical specifications and capabilities of all major world-wide CNT manufacturers) revealed the following status:

- The format and the parameters specified were not identical for any of the manufacturers. One manufacturer even stated that it did not make sense to specify any parameters, since the production process could not really be controlled.

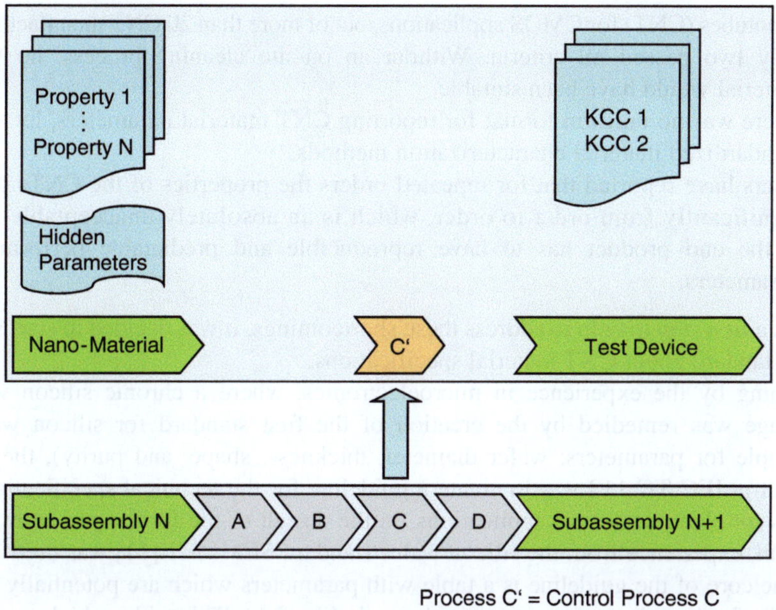

Fig. 5.9 Visualization of the role of "hidden" material parameters, which are difficult to characterize directly, but can be screened by a special product-related (pragmatic) control process C', which yield a KCC which allows to predict whether the nano material is suitable for the product application

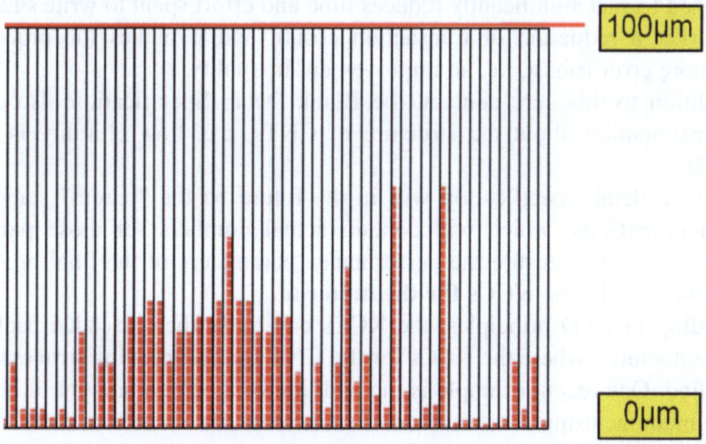

Fig. 5.10 Overview of the length of carbon nanotubes by different manufactures, as specified by the respective manufacturers. Each *bar* represents one company

- The capabilities varied widely among manufacturers, as an example the average length of CNTs is plotted per manufacturer (Fig. 5.10).
- It was reported at TC 113 Working Group 3 meeting in Gaithersburg, Maryland (USA) in Nov 2008 [8] that in a systematic screening process of carbon

nanotubes (CNT) for CMOS applications, out of more than 20 CNT manufacturers, only two passed all criteria. Without an on-site cleaning process, no CNT material would have been suitable.

- There was no uniform format for reporting CNT material parameters, let alone standardized material characterization methods.
- Users have reported that for repeated orders the properties of the CNTs differ significantly from order to order, which is an absolutely unacceptable state if the end product has to have reproducible and predictable performance parameters.

As a first step to help to address these shortcomings, it was decided to start work on a standard about CNT material specifications.

Going by the experience in microelectronics, where a chronic silicon wafer shortage was remedied by the creation of the first standard for silicon wafers (example for parameters: wafer diameter, thickness, shape, and purity), the first project in IEC TC 113 was to create a guideline for the technical specification of CNTs for electrotechnical applications. In the design of the format, structure, and content, experience from the silicon wafer standard SEMI M1 [11] was used.

The core of the guideline is a table with parameters which are potentially relevant for final products, an excerpt is shown in Fig. 5.11. This table, which contains the parameters, space for the upper and lower limit and the preferred measurement method has **no** concrete numbers in it. This concept is called a "**Blank Detail Specification**," and it provides a uniform format for the presentation and documentation of the relevant CNT material parameters. Experience shows that such a standardized format significantly reduces time and effort spent to write such specifications (i.e. a reduction of transaction costs), and it makes the procurement process more error robust, i.e. an improvement in quality.

In addition to this core content, the Blank Detail Specification also contains general information about the structure of CNTs, e.g. how chirality is defined (Fig. 5.12).

The blank detail specification will in the future be the "parent" standard for **detail specifications**, which will contain actual numbers for those parameters which are relevant to ensure the performance parameters of the final product are met. In other words, the **KCCs for the material**.

According to QFD principles, the KCCs can be further cascaded back to the CNT manufacturer, where the KCCs for the **CNT manufacturing process** have to be identified. One recent example is the addition of a certain amount of helium to the CVD manufacturing process in order to increase the fraction of metallic CNTs (as opposed to semiconducting CNTs) above 90% [12]. This is desirable e.g. for applications where the conductivity of the CNTs is used to make a non-conductive polymer matrix electrically conductive, e.g. in transparent films used for display or CNT additives in the Titanium dioxide based photovoltaic cells.

The last step before publication of the draft standard document is **validation** of the standard by asking manufacturers and users for their feedback. The result will be new draft, in which parameters which are not likely to be relevant in the foreseeable future are omitted, to make the document more "user-friendly".

5 Basic Specification Requirements

A basic specification is one that describes a commercially and technically appropriate single walled-carbon nanotube product having stable quality and parametric control. Single walled-carbon nanotubes produced to this specification shall be qualified through routine process checks (in the manufacturing process of the carbon nanotubes), demonstrating that the process is in a state of control.

The list of characteristics provided in the table 3 should be used as the basic specification requirement.

6 Recommended single-walled carbon nanotubes specification format

6.1 General Procurement Information

Table 2 - Format for general information

ITEM	INFORMATION	Date
General Specification Number		
Revision Level Part Number / Revision		
Growth Method	[ ] Laser ablation; [ ] High pressure carbon monoxide process; [ ] CVD; [ ] Arc synthesis; [ ] Combustion [ ] Other (specify):	
Functionalization (details to be provided)	Covalent [ ] non-covalent functionalization [ ] end / tip functionilazation [ ] side wall functionalization [ ]	
Dispersion Agent		
Dispersion Method		

6.2 Single-walled Carbon Nanotubes Characterization

6.2.1 General Characteristics

Table 3 - Format for general characteristics (after Saito et al, ref. 1)

ITEM		SPECIFICATION	Recommended Method (s)	Other MEASUREMENT METHODs
3-1	External Diametre	[ ] Nominal [ ] ± Tolerance [ ] nm	TEM	AFM, Fluoresecence, SEM; SPM; Raman;PL
3-2	Length	[ ] Nominal [ ] ± Tolerance [ ] μm	SEM	TEM; SPM; Raman;

Fig. 5.11 Excerpt from the committee draft of PT 62625, in which the concept of a Blank Detail Specification is illustrated

5.3.3.2 Example 2: IEC/TS 62607-2-1 Nanomanufacturing – Key Control Characteristics Part 2.1: Carbon Nanotube Materials – Film Resistance

One of the first applications of CNTs in electronic products will probably be transparent and flexible foils for flexible displays. A KCC for the product is the sheet resistivity. The project PT 62607 under the leadership of Ha Jin Lee from Korea is

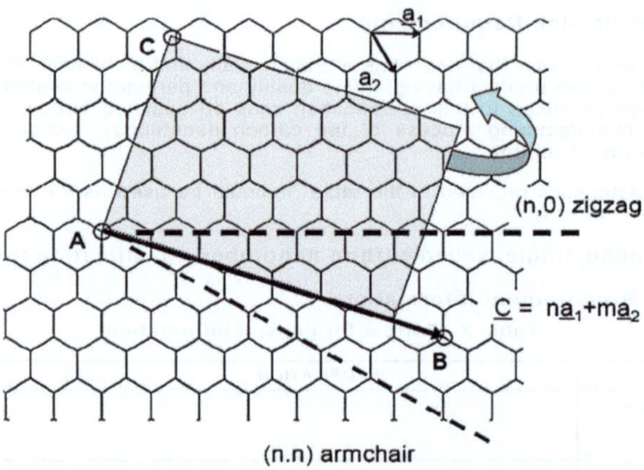

Fig. 5.12 Excerpt from the committee draft of PT 62625 from the part in which general information about CNT parameters is given, the excerpt explains the concept of chirality of carbon nanotubes in Fig. 5.1 of the draft document

developing a standardized preparation method for a CNT film from the material supplied by CNT vendors, which yields a value for the resistivity of the film, which can differ considerably from vendor to vendor. This "compound" parameter measured is influenced by a number of elementary materials parameters (such as the resistivity of the individual tubes, their average length, their surface properties, and their dispersion state). Measuring all these elementary material parameters and deriving a value for the sheet resistivity of a CNT thin film (ribbon) would be rather complex. Further, the most likely presence of some "hidden" parameter would render this task impossible. So the only feasible way is to use the pragmatic, product-related KCCs. The results of a trial-run (see Fig. 5.13) indicate that this indirect performance characterization method is relatively robust and relevant with respect to the application in flexible transparent films.

5.4 Future Developments

If the QM-based standardization strategy makes sense for nanoelectronic production, is it appropriate for the other nanotechnology products?

For any product that is supplied to safety and/or quality sensitive industry segments, such as automotive, aerospace or medical, the answer must be yes. Since the expectations for quantum leap improvements and breakthroughs through nanotechnology are high in those areas, it should be mandatory to use such a QM-centred strategy to ensure a timely and smooth industrialization. In this context, it should be noted that the application of QFD and other QM tools have reduced by 50% the development times in the automotive and other industries [13, 14] and significantly decreased the need for unplanned last minute changes after the start of production.

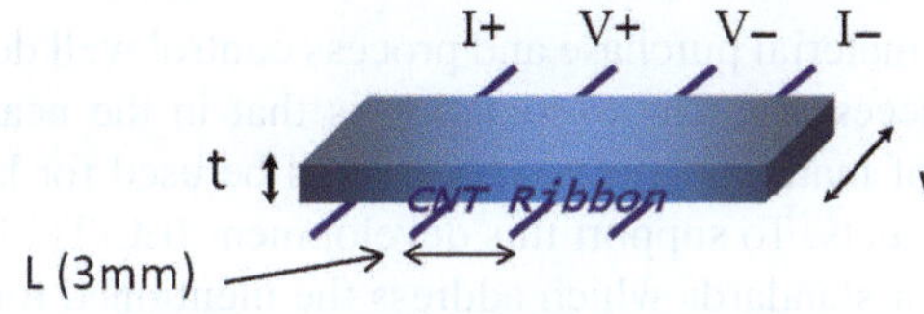

CNT	Units	1	2	3	4	5	Av
MWNT (A)	R (Ω)	19.03	27.27	27.04	20.83	20.38	
	ρs (Ω/sq.)	5.45	5.45	5.41	5.42	5.43	5.43±0.02
MWNT (B)	R (Ω)	2080	1920	1860	1680	1310	
	ρs (Ω/sq.)	693.3	672.0	620.0	616.0	679.5	656.17±35.7
MWNT (C)	R (Ω)	226.8	185.6	210.3	225.4	202.6	
	ρs (Ω/sq.)	83.92	89.09	92.53	78.89	83.07	85.50±5.35
SWNT (D)	R (Ω)	9.55	7	7.4	7.6	6.4	
	ρs (Ω/sq.)	1.43	1.40	1.53	1.52	1.79	1.53±0.15
SWNT (E)	R (Ω)	38.9	36.0	52.1	38.2	36.1	
	ρs (Ω/sq.)	14.00	12.60	18.24	16.43	14.44	15.1±2.21

Fig. 5.13 Schematic four contact measurement arrangement for the CNT ribbon (16) prepared from CNT samples from several suppliers. The results demonstrate that the method gives comparatively robust results and shows that there are significant differences between CNT suppliers (presentation given by the PT 62607 project leader given during a working group 3 meeting in April 2009) [16]

For commodity products which are neither quality nor safety sensitive there is, at first sight, no compelling need to follow the QM strategy. However, a well structured QM system, which in terms of how much is controlled via KCCs is adapted to the reduced requirements (less KCCs and wider specification limits), will still have economic advantages, as has been shown in a general study about the economic benefits of a ISO 9001-based QM system [15, 17].

Therefore, our conclusion is that also for the commodity sector, there will still be benefits if QM principles are applied to standardization for those product sectors, provided the effort to determine the KCCs is commensurate to the expected benefits. As in the case of electronic products, good alignment to existing standardization activities is absolutely mandatory.

Since the standardization strategy for the electronic products outlined in Sect. 5.3 is in its initial stages, and only partially implemented, it goes without saying that the initial concept will go through *several learning cycles* during the next few years. Hence the systematic of standardization proposed in Sect. 5.3 is not to be regarded as the final "truth," but "work in progress".

The actual structure of the standards related to nanomanufacturing is shown in Table 5.1. The basic idea is that for a mature high quality fabrication standards are required for material purchase, nanomanufacturing processes and fabrication equipment. Additionally for material purchase and process control well defined key control characteristics are necessary. The expectation is that in the near future a well defined small number of materials and processes will be used for large scale fabrication of electronic products. To support this development IEC/TC 113 developed a numbering scheme for standards which address the mentioned four aspects of nano-fabrication, well structured but open for addition of more materials, processes, KCCs and equipment. These can be added according to the requirement of the stakeholders of the industry.

5.5 Conclusions

One of the first questions answered in the first sections of this chapter is about **whether** anticipative standardization for the performance of nano-enabled products makes sense, was that anticipative standardization is by no means detrimental to innovation and technical progress in a key enabling technology, such as nanotechnology. It has been mentioned that the microelectronics industry, which undoubtedly is one of the most innovative industries, the almost ubiquitous use of standards has saved cost in the two digit percent range, and anticipative standardization was the facilitator for the industry transition from 200 to 300 mm wafer diameter.

In fact, in a more general macroeconomic context, a new picture is emerging. Among large industry segments with approximately more than $100 billion turnover, microelectronics is the segment with the largest growth rate, and this is mainly attributed to the high innovation rate. In line with this, a study by AFNOR [18], the French national standardization agency, the contribution of standards to economic growth is 1%, i.e. a considerable fraction of the economic growth figure of typically 3% in mature OECD economies (under normal circumstances) is due to standardization. Remarkably, a study of the economic impact of standards by DIN, the German national standardization agency, the economic impact of standards is larger than that of all patents.

So the *first conclusion* reached was: **Anticipative Standardization at an early stage makes sense**.

The lead principle in answering the second question, **how** to approach anticipative product performance related specification, was user centricity and how the specified performance of a nano-enabled product can be **ensured** rather than merely **tested**. The managed performance, reliability and durability (which can be regarded as a "long hand" description of the quality of nano-enabled electronic products) is implemented by quality and process management principles which have been used in quality and safety sensitive industries with great success for decades. The central idea is the implementation of the KCC concept, i.e. the identification of those material and process parameters which are key to ensuring that the technical specifications for the performance parameter of the final product are met. In other words, quality = performance cannot be inspected into a product, but is actively managed via the KCCs.

The KCC model has been applied to CNT specifications and to a test suitable for incoming inspection or in-line process control, other projects for all stages of the value adding chain are in their initial stages.

This KCC model can be applied not only to nanoelectronics but to all other areas of the manufacturing of nano-enabled products. The number of KCC parameters and the "allowed" limits for each parameter will, of course, depend on the application. An instructive example of what happens if there is no management of the critical process and material parameters and a stringent process and quality management, is the current situation in photovoltaics, which is a true mass production industry by now. PV module makers have been repeatedly confronted with "uncontrolled" changes of materials, e.g. the silver paste to print the front grid fingers. This can affect the soldering properties to such an extent that the soldering of wafers into strings does not work, whereas with the old type of paste there were no problems. In a similar vein, any change in the nanomaterials used for batteries can have a profound effect on the performance and durability of the final product, even if the change was made in the best intention. It is obvious that it is more difficult and expensive to correct such problems at a stage of mass production, rather than to set up a systematic KCC model to avoid such problems in the first place.

Going beyond nanotechnology, we propose that there should be a systematic study of all ISO and IEC standards related to the manufacturing of products whether there is improvement potential for those standards, and whether the KCC principle can be integrated.

References

1. http://www.nanotechproject.org/inventories/
2. Li, Y., Tan, B., Wu, Y.: Mesoporous Co3O4 nanowire arrays for lithium ion batteries with high capaciy and rate capability. Nanoletters **8**, 265–270 (2008)
3. ISO 9001:2008. ISO Standards Development Organisation, Geneva. http://www.iso.org
4. Bergholz, W., Weiss, B., Lee, C.: Benefits of standardization in the microelectronics industry and its implications for nanotechnology and other innovative industries. In: International Standardization as a Strategic Tool. Commended papers from the IEC Centenary Challenge, pp. 35–50 (2006)
5. Geleng, J.: Infineon Technologies. Private communication (2000)
6. ISO TS 16949:2009. ISO Standards Development Organisation, Geneva. http://www.iso.org
7. Cristiano, J., Liker, J., White III, C.: Customer-driven product development through quality function deployment in the U.S. and Japan. J Prod Innov Manage **17**, 286–308 (2000)
8. Winkler, R., Behnke, G.: Gate oxide quality related to bulk properties and its influence on DRAM device performance. In: Huff, H.R., Bergholz, W., Sumino, K. (eds.) Semiconductor Silicon 94, p. 673. The Electrochemical Society, Pennington, NJ (1994)
9. http://www.itrs.net/news.html
10. Segal, B.: Minutes of the IEC TC 113 Working Group 3 Meeting in Gaithersburg, Washington, DC, Nov 2008
11. Semiconductor Materials and Equipment International, San Jose, CA. http://www.semi.org
12. Harutyunyan, A.R., Chen, G., Paronyan, M., Pigos, E.M., Kuznetsov, O.A., Hewaparakrama, K., Min Kim, S., Zakharov, D., Stach, E.A., Sumanasekera, G.U.: Preferential growth of single-walled carbon nanotubes with metallic conductivity. Science **326**, 116–120 (2009)

13. Clausing, D.: Total Quality Development. American Society for Mechanical Engineers Press, New York (1994)
14. Slabey, W.R.: QFD: A basic primer – excerpts from the implementation manual for the three day QFD workshop. Transactions, second symposium on QFD, Novi, MI, 18–19 June 1990
15. Lo, C.K.Y., Yeung, A.C.L., Cheng, T.C.E.: Impact of ISO 9000 on time-based performance: an event study. Int J Humanit Soc Sci **1**, 35–40 (2009)
16. Lee, H.-J.: Private communication
17. Terziovski, M., Power, D., Sohal, A.S.: The longitudinal effects of the ISO 9000 certification process on business performance. Eur J Oper Res **146**, 580–595 (2003)
18. AFNOR: The Economic Impact of Standardization – Technological Change, Standards Growth in France. AFNOR, Paris (2009)

Chapter 6
Current Standardization Activities of Measurement and Characterization for Industrial Applications

Shingo Ichimura and Hidehiko Nonaka

6.1 Introduction

This chapter briefly describes current standardization activities for measurement and characterization of nanotechnology in various standardization organizations, with emphasis on the activity of ISO (International Organization for Standardization). Since the establishment of the U.S. National Nanotechnology Initiative (NNI) in 2001, both industrial and developing countries have accelerated investment for research and development (R&D) of nanotechnology [1]. In accordance with the increase in attention to nanotechnology worldwide, the interest in standardization for nanotechnology became prominent in 2004 in a trilateral framework involving the U.S., Europe, and Asia.

In Europe, the Technical Board (BT) of CEN, the European Committee for Standardization, established Working Group (WG) BT WG 166 on Nanotechnology in March 2004. Its major task was to analyze the need for standardization activities in the new area and to initiate relevant activities (see http://www.cen.eu). The task was later succeeded by new Technical Committee (TC) 352 on Nanotechnology, which was established in 2005. In the United States, ANSI (American National Standards Institute) established a NSP (Nanotechnology Standards Panel) in August 2004, to serve as the cross-sector coordinating body for facilitating the development of standards in the area of nanotechnology (see http://www.ansi.org). In Japan, the JSA (Japan Standards Association) also established a NSP in November 2004 to discuss and prepare a draft roadmap for nanotechnology standardization based on the request from METI (Ministry of Economy, Trade and Industry). In all of these activities, standardization of measurement and characterization for nanotechnology is a key issue.

S. Ichimura (✉)
National Institute of Advanced Industrial Science and Technology (AIST),
1-1-1, Umezono, Tsukuba 305-8568, Japan
e-mail: s.ichimura@aist.go.jp

V. Murashov and J. Howard (eds.), *Nanotechnology Standards*,
Nanostructure Science and Technology, DOI 10.1007/978-1-4419-7853-0_6,

Standardization of measurement and characterization for nanotechnology is critical not only to promote industrial application of nanotechnology, but to bring about social acceptance of nanotechnology. It is well known that there have been negative social responses toward nanotechnology [2, 3], leading to the call for adoption of precautionary actions for the handling of nano-materials [4]. This negative social response is based on several reports suggesting that nano-materials (nano-objects) including carbon nanotubes (CNTs) are harmful to human health and/or ecosystems [5–7]. However, precise investigations on the effects of nano-materials must be performed and reported based on common measurement protocols and reference materials. Standardization in measurement and characterization is, therefore, the key issue in the promotion of public awareness of the risks of nanotechnology.

Based on the aforementioned activities, ISO established TC 229 on Nanotechnologies in 2005. As of year-end 2009, ISO TC 229 has already convened nine general meetings (the ninth meeting was held in October 2009 in Tel Aviv, Israel) since its first meeting in November 2005 in London. At present, 33 national member bodies (i.e., countries) serve as P-members which actively contribute to the activity with voting rights, and 11 national member bodies serve as O-members (observers). It should be emphasized that not only are countries of chief industrial importance contributing to the activities of TC 229, but newly developing nations are as well.

The scope of ISO/TC 229 clearly states that it focuses on standardization in the field of nanotechnologies, which includes;

1. Understanding and control of matter and processes at the nanoscale, typically, but not exclusively, below 100 nm in one or more dimensions where the onset of size-dependent phenomena usually enables novel applications
2. Utilizing the properties of nanoscale materials that differ from the properties of individual atoms, molecules, and bulk matter, to create improved materials, devices and systems that exploit these new properties

The process of standardization in nanotechnology should reflect the difference of two major approaches of nanotechnology, these being the "top-down" approach and "bottom-up" approach. It is well known that the "top-down" approach is based on further advancement of current micro fabrication techniques. The approach aims to replace conventional industrial technologies with nanotechnologies, so that it might equally be called *evolution nanotechnology*. The annual target for R&D is often given quantitatively on a roadmap in this approach, as exemplified by ICT (Information and Communication Technology) and the electronics industries. With this approach, it is relatively easy to recognize when an existing standard should be revised or a new standard should be established. Thus, nanotechnology standardization relating to this approach might be called as "stand-by" type, and is mainly discussed in existing TCs of ISO rather than in TC 229. A good example of this type of standardization would be a set of nanoscale standards to be developed along International Technology Roadmap for Semiconductors (ITRS). Figure 6.1 schematically shows the direction of standards activities by existing standard organizations, and by new standards organization for nanotechnology (such as ISO/TC 229).

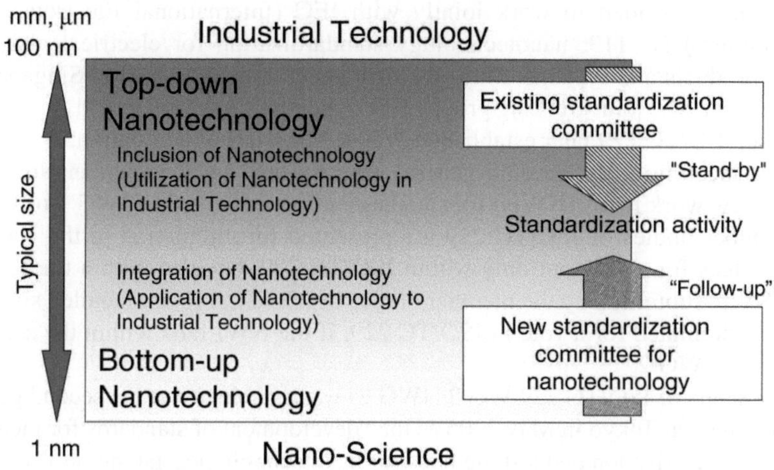

Fig. 6.1 Schematic view of the standardization activities relating to nanotechnology

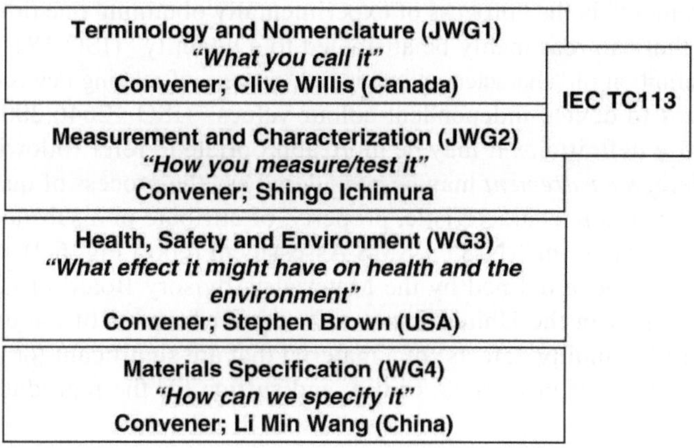

Fig. 6.2 Current structure of ISO/TC 229 and its relationship to IEC/TC 113

The "bottom-up" approach, based on assembling atomic level structures using mainly self-organization mechanisms, aims to open up a new phase in industrial technologies, which may be called *revolution nanotechnology*. One always has to await innovative applications after discovery and/or the assembling (creation) of novel nano-materials/structures. Therefore, standardization in this approach might be called as "follow-up" type, and it should have higher priority on the subjects relating to "what we call it", "how we measure it", and to "what effect it might have".

Figure 6.2 shows the structure of ISO/TC 229. It started with three WG structures beginning at the second meeting held at Tokyo in May 2006. The three WGs are WG 1, Terminology and nomenclature, WG 2, Measurement and characterization, and WG 3, Health, safety and environment. Among them, WG 1 and

WG 2 have decided to work jointly with IEC (International Electrotechnical Commission)/TC 113, nanotechnology standardization for electrical and electronic products and systems. Since the fifth general meeting held at Singapore in December 2007, joint working group (JWG) meetings have been held. In addition to these, ISO/TC 229 later established WG 4, Materials Specification. WG 4 held its first meeting at the seventh general meeting held in Shanghai in November 2008. New work items (NWIs) for standardization submitted to JWG 1 or JWG 2 by member bodies of ISO/TC 229 are presented for circulation to the National Committees for a vote not only within ISO/TC 229, but also within the IEC/TC 113. NWIs submitted by the member bodies of IEC/TC 113 are handled similarly and are circulated for a vote in ISO/TC 229, if the NWI falls within the scope of JWG 1 or JWG 2.

The scope of ISO/TC 229/WG 2 (JWG 2) was established at the second general meeting held at Tokyo in May 2006 as the "development of standards for measurement, characterization and test methods for nanotechnologies, taking into consideration needs for metrology and reference materials".

In the ISO concept data base, we can find 37 definitions for "measurement", and three definitions for "characterization". A typical example for the definition of "measurement" is the "process of experimentally obtaining one or more quantity values that can reasonably be attributed to a quantity" (ISO 18113-1:2009), while the definition of "characterization" is the "process of relating device-dependent colour values to device-independent colour values" (ISO 12646:2008). Instead of using those definitions, it may be more appropriate to refer following considerations. Here, *measurement* may be considered as "the process of quantitatively comparing a variable characteristic, property, or attribute of a substance, object, or system to some norm" (NIST USMS Assessment report Fig. 6.2), while *characterization* has been defined by the Materials Advisory Board of the National Research Council in the United States as "…those features of the composition and structure (including defects) of a material that are significant for a particular preparation, study of properties, or use, and suffice for the reproduction of the material" [8].

The first step for measurement and characterization standardization, therefore, might be different whether ISO/TC 229/WG 2 puts higher priority on *measurement* or if it puts higher priority on *characterization.* The prioritization of necessary *norms* (or standard units) for nanotechnology and preparation of the test methods for norms would be important in the former case, while prioritization of important *materials* in nanotechnology fields and preparation of the techniques and protocols for the characterization of the materials would be essential in the latter case. ISO/TC 229/WG 2 had discussed this point at the second ISO/TC 229 meeting held at Tokyo and decided to focus at first on the *characterization* of nano-materials, especially on CNTs.

In parallel with the discussion on the characterization of CNTs, ISO/TC 201/WG 2 established the Study Group (SG) on Strategy, led by Dr. Kamal Hossain (UK) and experts nominated from member bodies. The SG on strategy surveyed and discussed measurement and characterization needs for nanotechnology, based on the following objectives:

1. To develop measurement and characterization standards for use by industry in nanotechnology-based products
2. To work closely with all the ISO/TC 229 and IEC/TC 113 working groups in producing urgent standards of common interest by developing the necessary characterization, measurement and test standards
3. To ensure coordination with relevant work in other ISO/TCs, and TCs of other standards bodies developing measurement and characterization standards, and with OECD Committees, as appropriate
4. To promote the involvement of stakeholders in standardization activities and pre-normative research
5. To collect relevant inputs and formulate a systematic prioritization approach for standardization needs to support the development of an effective work programme for JWG 2

The SG submitted the report to ISO/TC 229/WG 2 and identified six key areas with higher priority together with possible themes for standardization based on data taken from the ISO/TC 229 questionnaire. The proposed areas with high priority are listed in Table 6.1 [9].

The following will briefly introduce current standardization activities for measurement and characterization of engineered nanomaterials (Sect. 6.2), and standardization activities for coatings/nanostructures (Sect. 6.3). Special focus will be placed on carbon nanomaterials (nanotubes) in Sect. 6.2 based on the activities of ISO/TC 229/WG 2, and on analytical techniques for coating and/or nanostructure measurement in Sect. 6.3, based on the activities of ISO/TC 201, surface chemical analysis. Standardization activities other than ISO/TC 229 and ISO/TC 201, such as ISO/TC 24/SC 4, IEC/TC 113 and IEEE will be briefly introduced (Sect. 6.4) together with activities of regional standards organizations such as CEN/TC 352 and ASTM E42 and E56.

Table 6.1 Priority areas for measurement and characterization of nanotechnology based on the report of the SG on strategy of ISO/TC229/WG 2

Priority area	
A	Standards for measurement and characterisation of carbon nanotubes and related structures
B	Standards for measurement and characterisation of engineered nanoparticles
C	Standards for measurement and characterisation of coatings
D	Standards for measurement and characterisation of nanostructured materials (composites and porous structures)
E	Standards for basic metrology at the nanoscale
F	Guidance for characterisation, specification and production of reference materials

6.2 Standardization for Measurement/Characterization of Engineered Nanomaterials Including Nanotubes (Activities of ISO/TC 229/WG 2 on Nanotechnology)

6.2.1 Representative Engineered Nano-materials

Various nano-materials have been fabricated and utilized with the aim of creating new functions. Table 6.2 lists representative examples of engineered nano-materials with the amount used based on a 2006 survey conducted in Japan [10]. Carbon black is the largest amount used at about 0.83 million ton, and is mainly used for automobile tires to increase durability. Silicon dioxide (silica) is the second largest amount used (about 13,500 ton) with its major application as an additive for silicone rubber film, fiber-reinforced plastic, and others. Titanium dioxide (about 1,250 ton) follows with major applications in the fields of cosmetics and toner. Table 6.3 summarizes estimated market sizes for selected nanomaterials in Europe as in 2006, which are available in NanoRoadSME site. Although the market sizes are diversely given either by weight or amount of money, all the market sizes are expected to grow quickly to form large markets by mid 2010s.

Table 6.2 Representative examples of engineered nano-materials and the used amount of them in Japan 2006

Carbon nanotubes (CNTs)	
Single-wall (SWCNTs)	0.1
Multiwall (MWCNTs)	60
Carbon nanofiber	60–70
Carbon black	8.3×10^5
Fullerenes	2
Dendrimers	50
Zinc oxide	480
Titanium dioxide	1.25×10^3
Silicon dioxide	1.35×10^4
	(unit; ton)

Table 6.3 Estimated market size for nanomaterials in Europe as in 2006 (quoted with modification from Fig. 3 in NanoRoadSME)

Years	2006 2007	2008 2009 2010	2011 2012 2013 2014
Nanomaterials	short term	middle term	long term
Carbon black		~9.6 million tons	
Carbon nanotubes	~700 million $	~3.6 billion $	~13 billion $
Hydrophobic fumed silica nanoparticles	~10 million $	~13 million $	~40 million $
Montmorillonite nanoclays (patelet)	~10 million $	~13 million $	~40 million $
Polymer with carbon nanoparticles/fillers (bulk)	~21 million $	~30 million $	~75 million $
Silicon carbide nanofibers	~150 tons/year	~1500 tons/year	~3000 tons/year
Titanium nanoparticles	~1500 tons/year	~3500 tons/year	~7500 tons/year

The amount of CNTs used is only 60 tons, and its major application, in the form of multiwall CNT (MWCNT), is for anti-static trays for the semiconductor industry. However, ISO/TC 229/WG 2 decided at the second ISO/TC 229 meeting held at Tokyo to place the highest priority on standardization of CNTs for the following reasons. After it was first discovered, using transmission electron microscopy (TEM), that carbon atoms can be aligned to form tubes [11], various substructures of CNT (single-wall, double-wall, and multiwall) and their homeomorphous forms such as graphene, nanohorns, and others have been identified. CNT can have both metallic and semi-conductive characteristics based on its chirality. Moreover, it becomes clear that CNTs have high tensile strength (100 times greater than iron), high electron mobility (1,000 times larger than conventional semiconductor materials), high electron emittance (100 times larger than conventional electron beam sources), high thermal conductivity (several times higher than diamond), high hydrogen absorbability (five times higher than metals), and low density (as half as aluminum) [12, 13].

In addition to the industrial importance, standardization for potential risk assessment of CNTs is quite important and urgent, as it was recently reported that CNTs have the possibility of inducing mesothelioma [6, 7]. The report from the SG on the Strategy mentioned above also suggested the urgent need for standardization of items relating to risk assessment as shown in Fig. 6.3. This need is based on the results of a questionnaire conducted by ISO/TC 229 asking for rankings of priority and urgency for the preparation of standards. The survey results for CNTs clearly stated that the item with highest priority and highest urgency is inhalation testing. It was also clear that other items such as toxicology testing, exposure determination and safe handling have high priority and urgency together with characterization needs for diameter distribution, sampling methods, length distribution, and chemical structure. It should be mentioned that guidance for the physicochemical characterization of

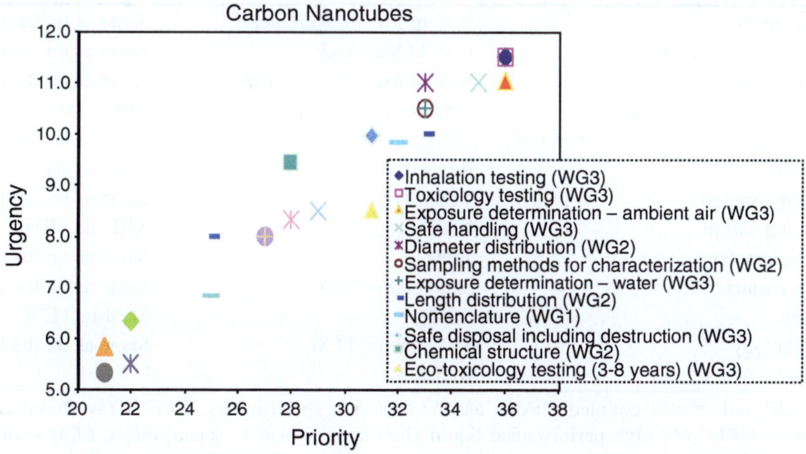

Fig. 6.3 Analysis of priorities for possible new work items relating to CNT [9]. Priorities are based the results of a questionnaire to ISO/TC 229 member bodies

engineered nanoscale materials for toxicological assessment is now under discussion between ISO/TC 229/WG 3 and WG 2 through the Joint Task Group on Measurement and Characterization for EHS Issues, led by Dr. Angela R. Hight Walker (USA) to help the activity of the OECD.

6.2.2 Standardization of MWCNT Characterization

Sophisticated synthesis methods have been investigated to improve purity and crystallinity (graphitization) of CNTs. With the improvement of the synthesis methods, measurement and characterization techniques for them have been concurrently developed and improved. Thus ISO/TC 229/WG 2 has surveyed measurement properties and possible measurement methods which have to be applied for the characterization of MWCNT by the questionnaire circulated for experts in member bodies. The questionnaire asked whether each measurement property was important and useful for suppliers or for users with appropriate quality control, and asked if a given method was already well-established for the characterization of MWCNTs.

Table 6.4 shows the results of the survey answered by 25 experts nominated from six countries. Among 15 measurement properties listed in the questionnaire, the experts responded with the highest importance for the 13 properties shown in Table 6.4: that is, the experts agreed the importance of purity control by measurement of ash content, metallic constituents, volatile content, polyaromatic hydrocarbon content, and carbon materials excluding MWCNTs. For physical and geometric

Table 6.4 Measurement properties and measurement methods to be applied for the characterization of MWCNT based on the questionnaire survey to ISO/TC 229/WG 2 experts

Property	Measurement method	Method in TR10929
Ash content	Weight loss measurement	Same as on the left
Metallic residual content	ICP-AES or XRF	Same as on the left
Volatile content	Weight loss measurement	Same as on the left
Polyaromatic hydrocarbons	Volume resistivity measurement[a]	HPLC-MS
Carbon materials excluding MWCNT	SEM and/or TEM	Same as on the left
Disorder	Raman[a]	Same as on the left
Burning property	TGA/DTA	Same as on the left
Stacking nature	XRD[a]	XRD or TEM
Inner diameter	TEM	Same as on the left
Outer diameter	SEM and/or TEM	Same as on the left
Length	SEM[a]	SEM or TEM
Morphology	SEM and/or TEM	Same as on the left
Surface	BET	[b]

ICP-AES inductively coupled plasma atomic emission spectrometry, *XRF* X-ray fluorescence analysis, *HPLC-MS* high performance liquid chromatograph-mass spectrometer, *SEM* scanning electron microscopy, *TEM* transmission electron microscopy, *TGA/DTA* thermogravimetric analysis/differential thermal analysis

[a]Method which some experts considered as it is not yet well-established (appropriate)

[b]Property which is not included in TR10929

property control of MWCNTs, the experts also agreed that measurements of disorder, burning properties, inner/outer diameter, length, and morphology are necessary. The experts replied that most of the proposed measurement methods such as ICP-AES (Inductively Coupled Plasma-Atomic Emission Spectroscopy), XRF (X-Ray Fluorescence analysis), SEM (Scanning Electron Microscopy), TEM, TGA/DTA (Thermogravimetric Analysis/Differential Thermal Analysis), and BET are already useful, but they suggested some of them were not yet sufficiently established.

Based on the survey results, JISC (Japanese Industrial Standard Association) submitted a new work item proposal for the characterization of MWCNT as a Technical Specification (TS), to be published as a prospective standard. The proposal was response to an urgent request for guidance to meet an identified need. A TS is usually reviewed 3 years after its publication in order to consider its conversion to an International Standard (IS) with additional information. In the proposal, which was accepted as TR (Technical Report) 10929 instead of TS by the voting of member bodies, JISC added one measurement property (moisture content by weight loss measurement), deleted one measurement property (surface area), and proposed alternative measurement methods for some of the measurement properties as listed in Table 6.4. In TR 10929, principle, experimental procedure, and expression of the experimental results are explained and prescribed for each set of measurement properties and measurement methods. The TR will be published after revision based on the accepted comments from member bodies.

In addition to the properties listed in Table 6.4 for the characterization of MWCNT, the property of bending ratio of MWCNT has been proposed as a new work item from KATS (Korean Agency for Technology and Standards). Since MWCNTs synthesized by Chemical Vapor Deposition (CVD) have static (permanent) bend points randomly distributed along their axis, physical and chemical properties of mass-produced MWCNTs are strongly dependent on the statistical distribution of mesoscopic shapes and sizes of the individual MWCNT particles that comprise the mass produced product. It is therefore crucially important to characterize the mesoscopic shapes of MWCNTs in order to obtain reproducible final properties for their use in composites and solutions as well as for EHS investigations. The proposal was approved as TS 11888, and its development is in progress under ISO/TC 229/WG 2 Project Group (PG) led by Dr. H. Sang Lee. In addition to ICP-AES for the measurement of metallic residual content shown in Table 6.4, JWG 2 just began a discussion on a standard for ICP-MS (Mass Spectrometry) under PG led by Dr. C. Chen (China). Table 6.5 shows the list of current ISO/TC 229/WG 2 items related to the characterization of MWCNTs.

Table 6.5 List of current projects in ISO/TC 229/JWG 2 for the characterization of MWCNTs

Document	New work item	Member body
TR 10929	Characterization of multiwall carbon nanotubes (MWCNTs) – collection of measurement methods	JISC
TS 11888	Determination of mesoscopic shape factors of multiwall carbon nanotubes (MWCNTs)	KATS
TS 13278	Determination of metal impurities in carbon nanotubes (CNTs) using inductively coupled plasma-mass spectroscopy (ICP-MS)	SAC

6.2.3 Standardization of SWCNT Characterization

For the standardization of SWCNT characterization, ANSI (American National Standards Institute)/USA proposed the classification of characterization levels. That is, it proposed to start with the analysis of purity and structural properties as level 1 characterization, followed by the analysis of electrical, magnetic, mechanical, optical properties, and others as level 2 through analysis of functional properties (level 3), and analysis of interaction with other materials such as bio molecules (level 4). In addition, ANSI proposed to choose morphology, length and diameter, tube type, and dispersability/solubility as major targets for structural properties in level 1 characterization, and proposed the adoption of five measurement methods for an initial screening step (Part A), six methods for more detailed analysis (Part B), and other six methods for additional analysis (Part C) as part of the level 1 characterization. The methods proposed as Part A, Part B, and Part C are listed in Table 6.6. Here, "x" mark means that the measurement method can be applied for the characterization of the property.

Major aspects of morphology characterization in Table 6.6 are the analysis of tube structure, bundle thickness, and orientations by SEM/EDX (Energy Dispersive X-ray analysis), and the analysis of wall structure, amorphous carbon, and metal catalyst coatings by TEM. For additional characterization, the following target items are established: analysis of oxidation/transition temperature (by TGA); surface area and pore size (by surface area measurement); chemical binding state (by XPS; X-ray Photoelectron Spectroscopy); functional group and volatile component (by FTIR; Fourier Transform Infrared Spectroscopy); crystallinity (by XRD; X-ray Diffraction); and chemical binding state and neighboring atom information (by EXAFS; Extended X-ray Absorption Fine Structure).

Through discussion at the third ISO/TC 229 meeting held at Seoul, Korea in December 2006, TC 229/WG 2 decided as the first step on the standardization to focus on the five measurement methods listed as Part A, and to add two additional measurement methods, near infrared photoluminescence absorption spectroscopy (NIR-PL) and evolved gas analysis-gas chromatography mass spectrometry (EGA-GCMS). The objective of NIR-PL is to provide a "measurement method for the determination of the chiral indices of semi-conducting SWCNTs in a sample and their relative integrated PL intensities", while EGA-GCMS aims to provide "guidelines for the characterization of volatile impurities in SWCNTs".

Table 6.7 lists the current programs related to the characterization of SWCNTs with the name of the member bodies proposing them. All projects aim to issue a TS as the first step. It should be emphasized that two projects have been carried forward under co-leadership from two member bodies, i.e., ANSI and JISC for PG1 (TEM) and ANSI and KATS for PG7 (TGA), based on agreement at the Seoul meeting. It was also agreed to share the preparation of one document with other member bodies and to prepare a final document (as an IS) as early as possible.

Table 6.6 List of measurement properties and measurement methods proposed by ANSI

Level 1: Purity and structural properties		Morphology	Purity	Length and diameter	Tube type	Dispersability/solubility	Additional
Part A: initial screening step	SEM/EDX	x	x	x		x	
	TEM	x	x	x			
	Raman		x	x	x		
	UV-Vis-NIR absorption		x	x	x	x	
	TGA		x				x
Part B: more detailed analysis	Fluorescence spectroscopy			x	x	x	
	Surface area measurement						x
	XPS		x				x
	AFM			x		x	
	FTIR						x
	ICP		x				
Part C: additional analysis	STM				x		
	XRD						x
	XRF		x				
	EXAFS						x
	E-beam diffraction				x		
	Light, X-ray, neutron diffraction			x		x	

XPS X-ray photoelectron spectroscopy, *AFM* atomic force microscopy, *FTIR* Fourier transform infrared spectroscopy, *STM* scanning tunneling microscopy, *XRD* X-ray diffraction, *EXAFS* extended X-ray absorption fine structure

Table 6.7 List of current projects in ISO/TC 229/JWG 2 for the characterization of SWCNTs

Document	Work item	Member body
TS 10797	Characterization of single-wall carbon nanotubes (SWCNTs) using transmission electron microscopy (TEM)	ANSI JISC
TS 10798	Characterization of single-wall carbon nanotubes (SWCNTs) using scanning electron microscopy (SEM) and energy dispersive X-ray analysis (EDXA)	ANSI
TS 10868	Characterization of single-wall carbon nanotubes (SWCNTs) using UV-Vis-NIR absorption spectroscopy	JISC
TS 10867	Characterization of single-wall carbon nanotubes (SWCNTs) using NIR-photoluminescence (NIR-PL) spectroscopy	JISC
TS 11251	Characterization of single-wall carbon nanotubes (SWCNTs) using evolved gas analysis-gas chromatograph mass spectrometry (EGA-GCMS)	JISC
TS 11308	Characterization of single-wall carbon nanotubes (SWCNTs) using thermogravimetric analysis (TGA)	ANSI KATS
TS 10812	Characterization of single-wall carbon nanotubes (SWCNTs) using Raman spectroscopy	ANSI

Table 6.8 Relation of measurement method and purity information given by the method

Project	Method	Target of purity analysis
TS 10798	SEM/EDX	Non-carbon impurities
TS 10797	TEM	Tube surface cleanliness
TS 10812	Raman	Nanotube and nonnanotube carbon
TS 10868	UV-Vis-NIR absorption	Carbonaceous content (quantitative)
TS 11308	TGA	Non-carbon content (quantitative)
	XPS	Elemental composition (surface)
TS 13278	ICP	Elemental composition (quantitative)
	XRF	Elemental composition (quantitative and non-destructive)
TS 10867	NIR-PL	Relative mass concentrations of semi-conducting SWCNTs
TS 11251	EGA-GCMS	Volatile impurities (qualitative, and quantitative with weight loss measurement)

It is noteworthy that in Table 6.6 multiple measurement methods are considered to be applied for a specific property such as purity, length and diameter, etc. In the case of purity, for example, the following information is expected to be given by each measurement method (see Table 6.8).

The features expected to be obtained by each method in the purity assessment of SWCNTs are briefly summarized below based on the current working draft of each project. It should be noted that parts of the content will likely be modified based on discussion of experts before the final publication as a TS.

SEM/EDX (TS 10798), especially EDX analysis, can be applied to determine the elemental composition of non-carbonaceous impurities in SWCNTs. It has good sensitivity to impurities such as residual catalysts, surfactants, and acid functionalized products. It is typically used to generate qualitative data, and in some cases used to calculate semi-quantitative data if advanced software routines are utilized. It is mainly used to provide an average composition, so that dedicated TEM/EDX systems have to be used if, for example, the identification of a catalyst particle in CNT material is necessary.

TEM (TS 10797) can be applied for qualitative visual estimation of the purity of SWCNT. Impurities such as (metal) catalyst residues, and other typical by-products such as multiwall nanotubes, carbon nanofibres, fullerenes, amorphous carbon and graphite onions can also be assessed by visual and instrumental evaluation. Inorganic impurities such as metals, metal oxide or carbides, as well as heteroatoms such as N, S or Cl, can be analyzed spectroscopically by a combination of EDX and EELS (electron energy loss spectroscopy).

UV-Vis-NIR absorption spectroscopy (TS 10868) can be applied to measure relative purity, i.e., the content of SWCNT in the total carbonaceous content in the sample, from optical absorption peak area. Both specific absorptions of SWCNT originating from interband transitions, which are typically observed in the Vis-NIR region, and featureless background formed by Π-plasmon absorption of SWCNT and carbonaceous impurities are used for the analysis. Results are only qualitative because of many factors such as statistical uncertainty associated with the estimation of peak area intensity by linear baseline subtraction.

TGA (TS 11308) provides a quantitative measure of the non-carbon impurity (e.g., metal catalyst particles) level in SWCNT material, leading to an estimation of net fraction (weight percentage) of SWCNT within a given sample. It also allows quality assessment of the SWCNT material by providing residual weight and oxidation temperatures.

NIR-PL (TS 10867) can be applied to estimate relative mass concentration of semi-conducting SWCNTs in a sample from measured integrated PL intensities and knowledge of their cross-sections.

EGA-GCMS (TS 11251) gives qualitative information on volatile impurities in SWCNT by comparing measured mass spectra with mass spectral databases for standard compounds. It can also give quantitative information of evolved gas components in a SWCNT sample by weight loss measurement using a microbalance before and after heating for mass analysis.

The working draft for Raman (TS 10812) and ICP-MS (TS13278) has not yet been prepared.

6.2.4 Necessity of Standardization for the Characterization of Other Engineered Nanomaterials

Figure 6.4 shows the current roadmap of ISO/TC 229/WG 2. As explained above, TC 229/WG 2 started its activity by focusing on carbon nano-materials. Therefore, a possible next step would be the measurement and characterization of fullerenes and carbon black since they are selected as target materials for OECD working programs in addition to SWCNT and MWCNT as shown in Table 6.9.

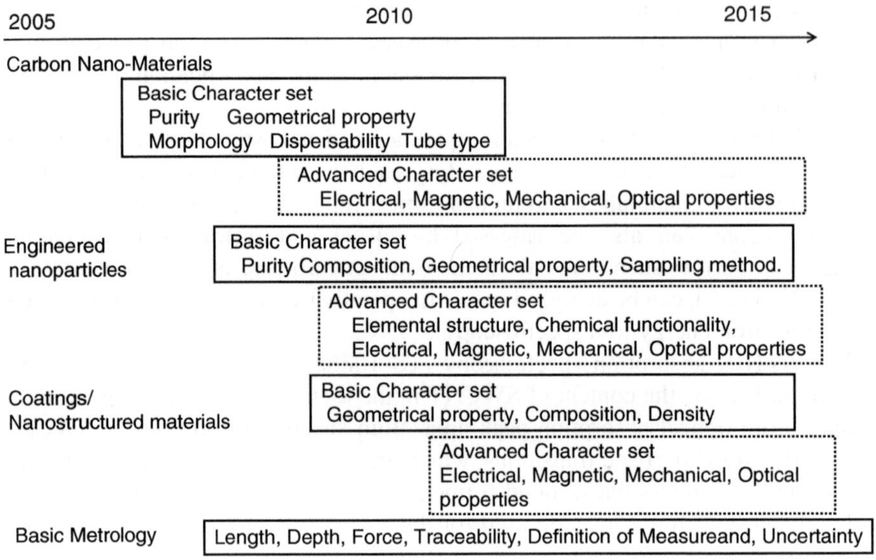

Fig. 6.4 Draft roadmap and future plan of WG 2 based on the outline strategy for ISO/TC 229/WG 2 [20]

Table 6.9 Engineered nano-materials selected as OECD sponsorship arrangements

Fullerene (C60)	Aluminum oxide
SWCNT	Cerium oxide
MWCNT	Zinc oxide
Silver nanoparticles	Silicon dioxide
Iron nanoparticles	Polystyrene
Carbon black	Dendrimers
Titanium dioxide	Nanocray

It is now well known that carbon atoms are aligned to form a soccer-ball-like sphere known as a Fullerene. Its existence was theoretically predicted in 1970 [14] and was actually discovered in 1985 [15]. From the first discovery, various derivative structures have been synthesized. Fullerenes and various fullerene derivatives are all rigid spherical molecules, are soluble in organic solvents, and have the acceptability of metal encapsulation [16]. Thus their applications as electron acceptors, and as strong light absorbers are expected to include superconductivity through doping of metallic elements.

One good example of fullerene applications is that they are good candidates as a primary material for proton transport membrane in fuel cells [17]. Since many polar functional groups can be introduced on the compact surfaces of fullerenes, protons can be transferred without the presence of water. This means that a fuel cell with a fullerene membrane can be used at temperatures below 0°C or at higher temperatures such as 120°C and beyond [18]. Another fullerene application is as a key material in organic semiconductor devices. It is already known that fullerene film has excellent n-type semiconductor properties, with electron mobility comparable to that of amorphous silicon, especially when it is prepared by deposition in ultra-high vacuum conditions. A high quality crystalline thin film by simple a spin coating method has been developed by synthesizing new fullerene derivatives such as C60-fused pyrrolidine-meta-C12 phenyl (C60MC12), and by incorporating an alkyl chain to fullerenes [19]. Since both n- and p-type organic semiconductors with high electron mobility are obtained by the coating method, it will accelerate the practical application of fullerenes to small-sized organic electron circuits.

As to standardization of fullerene characterization, a survey of TC 229/WG 2 experts was conducted, like that for CNT characterization, shown in Table 6.4. The survey results suggested that the characterization of fullerene properties such as Fullerene composition (by liquid chromatography), thermal properties (by TGA/differential thermal analysis), surface area (by BET), residual solvent (by gas chromatography), and metal impurity (by ICP-AES, and/or atomic absorption spectroscopy) are important and possible. The survey also suggested that the analysis of particle diameter, particle size distribution, and pore size distribution are important issues, although the measurement methods for the analysis of those properties had not yet established. Standardization of fullerene characterization has not yet started in TC 229/WG 2.

Measurement and characterization of oxide nano-materials (titanium dioxide, aluminum oxide, cerium oxide, zinc oxide, and silicon dioxide) and metallic nanoparticles (silver nanoparticles and iron nanoparticles) are also important targets. They have also been selected as target materials for OECD working programs (Table 6.9) and some of them have larger production amount than CNTs as shown in Table 6.2.

Regarding characterization of engineered nano-particles, ISO/TC 229/WG 2 has already started the project of ISO 12025, “General framework for determining nanoparticle content in nanomaterials by generation of aerosols”. The purpose of the project is “to measure the number of particles in the size range from approximately 1 to 100 nm”, which are “generated by a defined treatment procedure of a defined nanomaterial sample” considering that “the primary nanoparticles of the nanomaterial as received before testing have not been significantly altered by sample preparation and testing” and therefore “the generated aerosol should be representative of the releasable nanoparticle content”. Since the generation of aerosol and its characterization has close relation with ISO/TC 24/SC 4 on sizing by methods other than sieving, the project is in progress under the collaboration with that SC.

It should be emphasized that the SG on metrology, which is under the purview of ISO/TC 229/WG 2, has discussed a metrological checklist which any proposer of a new work item must take into account in order to improve the quality of the submitted document. Table 6.10 shows the outline of the metrological check-list [20].

In this chapter the ISO standardization of measurement and characterization of engineered nanomaterials themselves has been described, but the functions and the performance of nanomaterials may also depend strongly on their surfaces and interfaces. The next Sect. 6.3 describes the standardization of measurement and characterization of surface and interface structures in view of surface chemical analysis.

Table 6.10 Outline of the metrological check-list prepared by the Study Group on Metrology [20]

1	Has the system/body/substance that will be subjected to the measurement procedure, clearly been described, including its state?
2	Is the definition of the system/body/substance not unnecessarily restrictive?
3	Is the measurand clearly described?
4	Has it been clearly indicated whether the measurand is operationally or method-defined, or whether the measurand is an intrinsic, structurally defined property?
5	Is the measurement unit defined? Are the tools require to obtain metrological traceability available?
6	Has the method already been validated in one or more laboratories?
7	Are any quality control tools available to enable the demonstration of a laboratory's proficiency with the test method?
8	Have the results of measurements using the proposed method already been published in peer-reviewed journals by several laboratories?
9	Is the instrumentation required to perform the test widely available?
10	Does the document propose a measurement uncertainty budget?

6.3 Standardization of Analytical Techniques for Nanocoating/ Structure Measurement (Activities of ISO/TC 201 on Surface Chemical Analysis)

6.3.1 Standardization in ISO/TC 201 for Surface Chemical Analysis as a Tool to Characterize Surfaces and Interfaces of Nano-coating/Structure

Currently, analyses of material surface and interface are essential not only for R&D of industrial products, but evaluation of the quality and performance of products themselves. This is because the functions and the performance of these products largely depend on their surfaces and interfaces as typically in the case of silicon devices where the interface conditions of different material layers determine the performance of the devices. However, because there is relatively little history on surface chemical analysis methods and instruments, international rules to utilize them correctly to the evaluation are essential. As such, ISO/TC 201 was established in 1992 to develop international standards for surface chemical analysis (SCA), with an aim toward "standardization in the field of surface chemical analysis in which beams of electrons, ions, neutral atoms or molecules, or photons are incident on the specimen material and scattered or emitted electrons, ions, neutral atoms or molecules, or photons are detected," noting that, "with current techniques of surface chemical analysis, compositional information is obtained for regions close to a surface (generally within 20 nm) and composition-versus-depth information is obtained with surface analytical techniques as surface layers are removed."

Since its inception, nine major technical topics have been addressed in ISO/TC 201 activities – eight of them were delegated to subcommittees (SC's) as shown in Table 6.11, with their scopes together with SC 9 and WG 3 which were, as later mentioned, established in 2004 and 2008 respectively.

Table 6.11 Structure of ISO/TC 201 (2009)

SC/WG	Titles	Scopes
SC 1	Terminology	Standardization of the definitions of terms used in surface chemical analysis
SC 2	General procedures	Standardization of the procedures common to two or more SC's of ISO/TC 201, such as specimen preparation and handling, specification and preparation of reference materials, and methods of reporting results
SC 3	Data management and treatment	Standardization of the design of data bases, for the transfer of data between instruments, and for specifying the properties of algorithms used for surface chemical analysis
SC 4	Depth profiling	Standardization of methods for instrument specification, instrument calibration, instrument operation, data acquisition, and data processing used to determine composition versus depth with surface analytical techniques
SC 5	Auger electron spectroscopy (AES)	Standardization of methods for instrument specification, instrument calibration, instrument operation, data acquisition, data processing, qualitative analysis, and quantitative analysis in the use of Auger electron spectroscopy for surface chemical analysis
SC 6	Secondary ion mass spectrometry (SIMS)	Standardization of methods for instrument specification, instrument calibration, instrument operation, data acquisition, data processing, qualitative analysis, and quantitative analysis in the use of secondary ion mass spectrometry, sputtered neutral mass spectrometry, and fast atom bombardment mass spectrometry for surface chemical analysis
SC 7	X-ray photoelectron spectroscopy (XPS)	Standardization of methods for instrument specification, instrument calibration, instrument operation, data acquisition, data processing, qualitative analysis, and quantitative analysis in the use of photoelectron spectroscopy with X-ray and other photon sources for surface chemical analysis
SC 8	Glow discharge spectroscopy (GDS)	Standardization of methods for instrument specification, instrument operation, data acquisition, data processing, qualitative analysis, and quantitative analysis in the use of glow discharge optical emission spectroscopy and glow discharge mass spectrometry for surface chemical analysis
SC 9	Scanning probe microscope (SPM)	Standardization of methods for instrument specification, instrument calibration, instrument operation, data acquisition, data processing, qualitative analysis, and quantitative analysis in the use of scanning probe microscopy for surface chemical analysis
TC 201/ WG 2	Total reflection X-ray fluorescence spectroscopy (TXRF)	Standardization of methods for instrument specification, instrument calibration, instrument operation, data acquisition, data processing, qualitative analysis, and quantitative analysis in the use of total reflection X-ray fluorescence spectroscopy for surface chemical analysis
TC 201/ WG 3	X-ray reflectivity (XRR)	Standardization of methods for instrument specification, instrument calibration, instrument operation, data acquisition, data processing, qualitative analysis, and quantitative analysis in the use of X-ray reflectivity for surface chemical analysis

As recent development in nanotechnology has put fabrication and control of a wide variety of nanostructured materials into practical use, requirements have arisen for surface chemical analysis to analyze and characterize such nanostructured materials. In order to address the requirement, ISO/TC 201 modified its scope in 2005 in which surface chemical analysis was re-defined to include "techniques in which probes are scanned over the surface and surface-related signals are detected" with a note saying, "with current techniques of surface chemical analysis, analytical information is obtained for regions close to a surface (generally within 20 nm) and analytical information-versus-depth data are obtained with surface analytical techniques over greater depths." Because scanning electron microscopy had already been within the scope of ISO/TC 202 (microbeam analysis), it was excluded from the modified scope.

At the same time the scope modification took place, ISO/TC 201/SC 9, scanning probe microscopy (SPM), was established for standardization of methods for instrument specification, instrument calibration, instrument operation, data acquisition, data processing, qualitative analysis, and quantitative analysis in the use of scanning probe microscopy for surface chemical analysis. The first SC 9 meeting was held in 2004 in Jeju, Republic of Korea, the home country of both the SC 9 chair and secretary. The issues discussed in the first meeting were: (1) methodology for traceable calibration of length scales using a traceable artifact; (2) guidelines for the determination of experimental parameters for Atomic Force Microscope (AFM) by choosing suitable artifacts as indicators; (3) specification for cantilevers for SPM in the measuring mode including their dimensions and physical properties; (4) instrument specification for SPM for compatibility of data obtained by different instruments; and (5) guidelines for the determination of experimental parameters for Near-field Scanning Optical Microscopy (NSOM), including the probe shape and the gap control. Five study groups (SG's) were organized in SC 9 to discuss the issues. The SG's were then elevated to working groups (WG's) where these issues have been continuously discussed. In SC 9, many standardization projects related to SPM are now under development within six WG's for specific topics (see Table 6.12). Among them, the projects which have been registered by the ISO Central Secretariat are listed in Table 6.13. The earliest publishing date of new ISO standards from SC 9 is expected to be 2011.

Table 6.12 WG's of SC 9 (Jan. 2010)

WG	Titles
1	Use of NSOM/SNOM
2	Effects of measurement conditions
3	Basic dimensional SPM calibration of SPMs
4	Application-oriented dimensional SPM calibrations
5	Calibration of probes
6	Guideline and reference material for electrical SPM (ESPM) such as EFM, SCM, KFM and SSRM for 2D-dopant imaging and other purposes

Table 6.13 ISO projects under development by SC 9

Reg. no.	Titles
27911	SCA – SPM – definition and calibration of lateral resolution of a near-field optical microscope
11039	Standards on the definition and measurement methods of drift rates of SPMs
11952	Guideline for the determination of geometrical quantities using SPM – calibration of measuring systems
11939	Standards on the measurement of angle between an AFM tip and surface and its certified reference material
11775	SCA – SPM – determination of cantilever normal spring constants
13095	SCA – procedure for in situ characterization of AFM probes used for nanostructure measurement
13096	SCA – SPM – guide to describe AFM probe properties
13083	SCA – standards on the definition and calibration of spatial resolution of electrical scanning probe microscopes (ESPMs) such as SSRM and SCM for 2D-dopant imaging and other purposes

6.3.2 Standardization of SPM in ISO/TC 201/SC 9

As seen from Table 6.13, the current projects within SC 9 are proposed for establishment of a basic standard system to determine SPM itself as a traceable tool. Basic topics for standardization include terminology, description of the instrumental components, and guidelines. Among the ongoing projects, we introduce briefly an approved new work item (AWI) 13095, "procedure for in situ characterization of AFM probes used for nanostructure measurement" as a typical case of the standardization of instrument calibration. This AWI, once established as an international standard, will specify a method for characterizing the shape of AFM probes to reduce the uncertainty of AFM measurement of nanostructures and/or nanomaterials. The need of precise shape and size measurements of nanostructures with AFM is quite obvious and has been verified by the results of a survey on key factors and sizes for the realization of new functions by using nanotechnology, conducted by the JSA in 2005 [21]. Figure 6.5 shows the results suggesting that AFM would be a powerful measurement tool for satisfying the need for shape and size measurements.

However, it is well known that in AFM measurements the operator should always be aware of observing artifacts created by the shapes of the probe, which is inappropriate for measuring the target nanomaterials. Figure 6.6 shows two specific cases demonstrating how the difference between the measured value of the width or height of the nanostructure and the real value is related to the probe shape used for the AFM observation. When two protrusive nanostructures with a narrow gap between them are measured, the depth measured between the two structures (H_m) is shallower than the height of the protrusion (H_o), as is shown schematically in Fig. 6.6a. On the other hand, in the measurement of an isolated protrusive nanostructure, the measured width (W_m) is wider than the width of the original structure (W_o), as is shown in Fig. 6.6b.

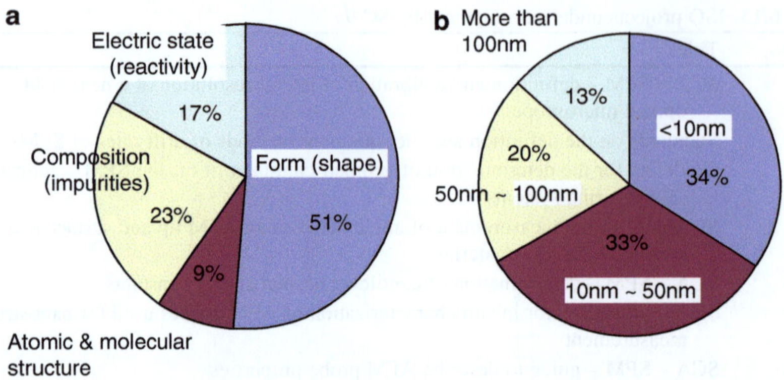

Fig. 6.5 Results of the questionnaire survey on (**a**) key factor and (**b**) key size to realize new function by nanotechnology. The analysis bases on the answers from 27 Japanese companies replied to the questionnaire survey [21]. Reproduced from [21] by permission from an IOP Publishing Ltd

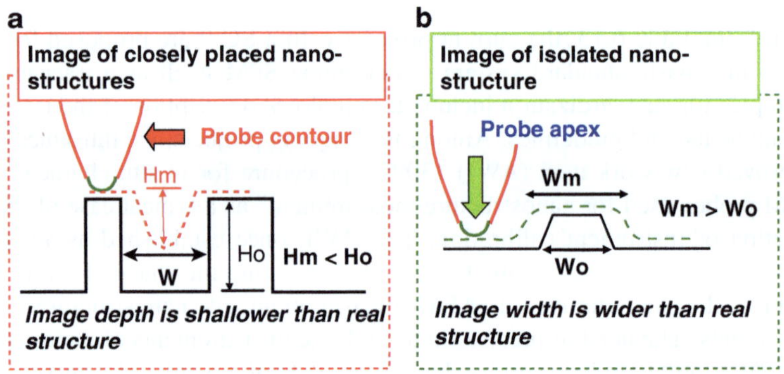

Fig. 6.6 Schematic illustration of AFM measurement results for (**a**) closely placed two nano-structures and (**b**) an isolated nano-structure. Reproduced from [21] by permission from an IOP Publishing Ltd

The work item shows the way to characterize the shape of an AFM probe as defined as Fig. 6.7 by using a reference sample with a comb-shape structure of 3–100 nm dimensions (for a typical design of the reference sample, see Fig. 6.7). In the annex, the AWI gives an example of sample fabrication using an GaAs/InGaP superlattices [22]. The apparent probe-shape characteristic can be obtained by imaging the trench structure of the reference sample as shown in Fig. 6.8. The dashed line represents a trace of the tip apex, which is obtained by the AFM measurement of a structure with rectangular hollows. The tip shapes at different positions of AFM measurement are shown with dotted lines. The probe shape characteristics can also be obtained by using the narrow-ridge structure of sample as shown in Fig. 6.9.

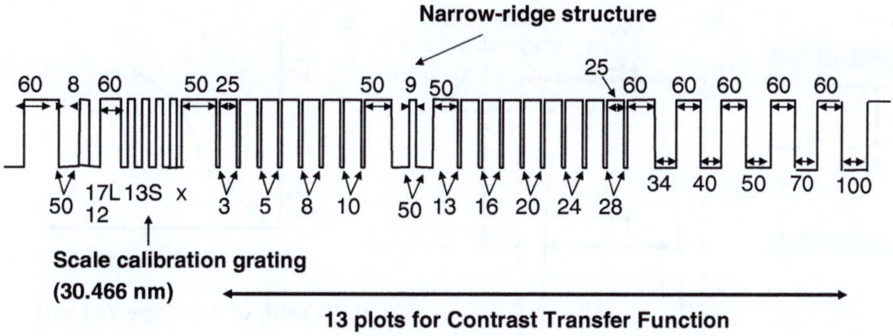

Fig. 6.7 Typical design of the reference sample with a comb-shape structure and a narrow-ridge structure

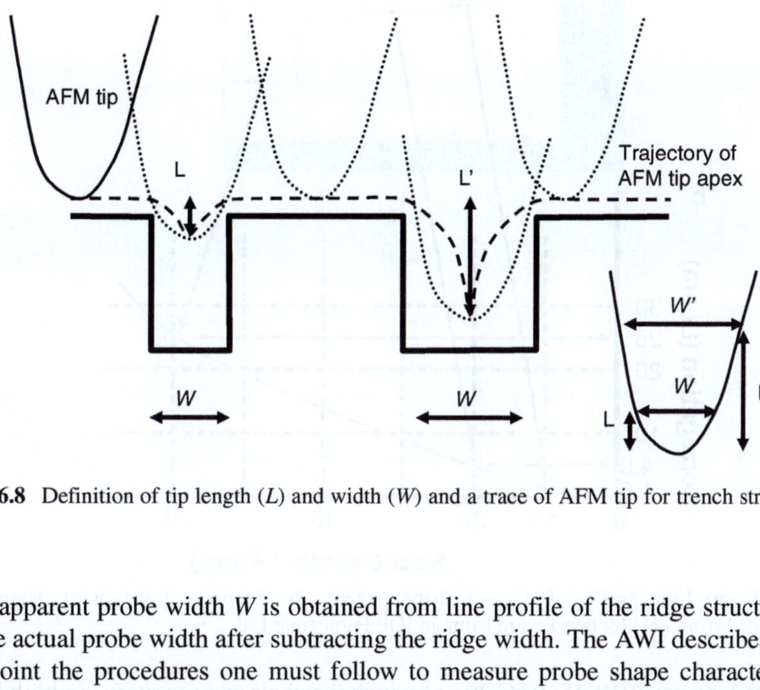

Fig. 6.8 Definition of tip length (*L*) and width (*W*) and a trace of AFM tip for trench structures

The apparent probe width *W* is obtained from line profile of the ridge structure. *W* is the actual probe width after subtracting the ridge width. The AWI describes point by point the procedures one must follow to measure probe shape characteristics using either trench structures or narrow-ridge structures. The average difference between the two probe-shape characteristics obtained using the trenches and the ridge is the order of difference between the actual and apparent probe shape. Figure 6.10 shows an example of line profile of a comb-shaped pattern of a reference sample obtained from an AFM image of the sample and the response function of the probe as a relation between the probe length and width estimated from the line profile. The aspect ratios of the AFM probe (*W*/*L*) can be estimated from the response function.

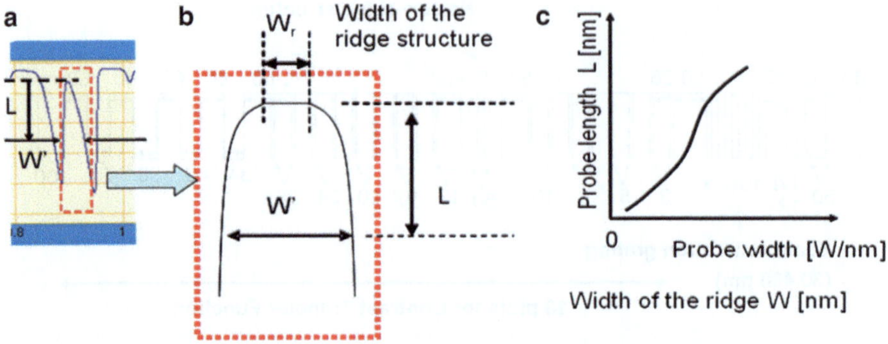

Fig. 6.9 Probe shape characteristic obtained using a narrow-ridge structure

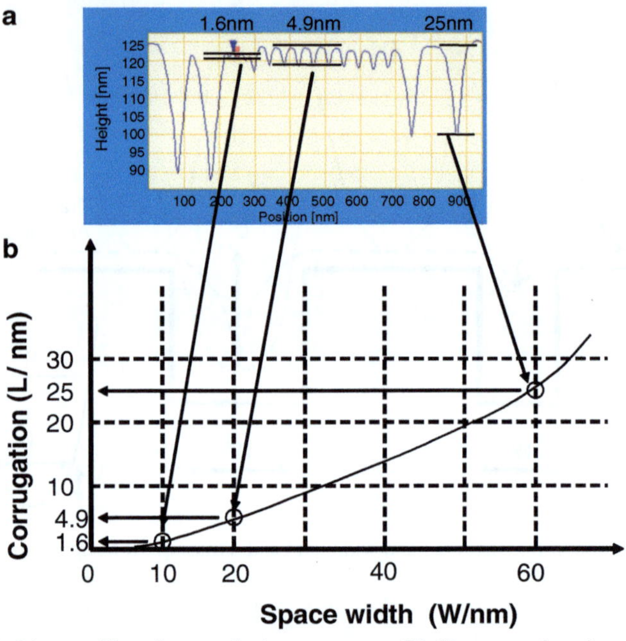

Fig. 6.10 (**a**) Line profile of a comb-shape pattern. (**b**) Response function of the probe. Reproduced from [21] by permission from an IOP Publishing Ltd

As of January 2010, ISO/TC 201 of which the chair and secretary are both from Japan, consists of ten P(participant)-members (Australia, Austria, China, France, Hungary, Japan, Republic of Korea, Russian Federation, USA, and UK) and 18 O(observer)-members (Brazil, Egypt, Finland, Germany, Hong Kong China, India, Ireland, Italy, Malaysia, Mongolia, Philippines, Poland, Romania, Singapore, Slovenia, Sweden, Switzerland, and Turkey), with organization liaisons with ISO/TC 202, IUPAC (International Union of Pure and Applied Chemistry), IUVSTA (International Union for Vacuum Science and Technique, and Applications), and

VAMAS (Versailles Project on Advanced Materials and Standards). The structure of ISO/TC 201 now consists of nine SC's and two in-line WG's.

WG 3 was established most recently in 2008 for standardization of subjects related to X-ray reflectometry. The technique is applicable to evaluation of thickness, density and interface width in the range 2–200 nm with a high accuracy even in multi-layer systems. Thus, standardization on XRR is not only within the scope of ISO/TC 201, but will provide clear guidelines to the technique for the application to the measurement and characterization of nanostructures. Currently, several new work item proposals by WG 3 for instrumental requirements and data acquisition are under preparation.

Since the first meeting in 1992 in Tokyo, 18 plenary meetings have been held, the last one of which was held November, 2009 in San Francisco, USA. At the San Francisco meeting, a pre-meeting of SC 9 was held in order to save time at the plenary meeting for discussion of many project issues from each SC 9/WG's. During the plenary meeting, a report of the study group on additional new work areas was made to suggest taking new analytical methods such as scanning laser confocal microscopy in its focus to characterize nanostructured materials, and there was a resolution to request the liaison officer of ISO/TC 201 for ISO/TC 229 to report the activities of ISO/TC 229 twice a year, though the official liaison with ISO/TC 229 had not yet been fully agreed upon.

6.3.3 Potential Use of International Standards Published by ISO/TC 201 for Nano-coating/Structure Characterization

By 2009 ISO/TC 201 developed 43 international standards including amendments of published standards, Technical Reports (TR), and Technical Specifications (TS) as listed in Table 6.13 ordered by the SC that developed them.

Although all of the standards in Table 6.14 are for uses related to surface chemical analysis, some potentially cover measurements at the nanoscale and hence can be utilized for the measurement/characterization of nanostructured materials. A good example is ISO 23812 published in 2009 by SC 6 for SIMS. The title of the standard is "method for depth calibration for silicon using multiple delta-layer reference materials" and standardizes a method for depth calibration of silicon using multiple delta-layer reference materials. Generally speaking, apart from its destructive nature, SIMS is a powerful method for measuring depth profiles of dopants in silicon, but as miniaturization of silicon devices has reached a manufacturing process scale comparable to that of nanostructures, a standardized method for determining dopant depth profiles in a shallow region, less than 50 nm from the surface, has been required. In such a shallow region, the accumulation of implanted primary ion species, oxygen or cesium, induces a sputtering rate change and a significant profile shift occurs when a uniform sputtering rate is assumed for depth calibration. To calibrate the depth scale in such a shallow region, it is proposed to use multiple

Table 6.14 International standards established by ISO/TC 201 by 2009

Ref.	Title
ISO 14606:2000	Surface chemical analysis (SCA) – sputter depth profiling – optimization using layered systems as reference materials
ISO 14706:2000	SCA – determination of surface elemental contamination on silicon wafers by total reflection X-ray fluorescence (TXRF) spectroscopy
ISO 14975:2000	SCA – information formats
ISO 14976:1998	SCA – data transfer format
ISO/TR 15969:2001	SCA – depth profiling – measurement of sputtered depth
ISO/PRF TR 16268	SCA – proposed procedure for certifying the retained areic dose in a working reference material produced by ion implantation
ISO 17331:2004	SCA – chemical methods for the collection of elements from the surface of silicon-wafer working reference materials and their determination by TXRF spectroscopy
ISO 17331:2004/DAmd 1	
ISO 18115:2001	SCA – vocabulary
ISO 18115:2001/DAmd1	
ISO 18115:2001/DAmd2	
ISO 18116:2005	SCA – guidelines for preparation and mounting of specimens for analysis
ISO 18117:2009	SCA – handling of specimens prior to analysis
ISO 22048:2004	SCA – information format for static secondary ion mass spectrometry (SIMS)
ISO/TR 22335:2007	SCA – depth profiling – measurement of sputtering rate: mesh-replica method using a mechanical stylus profilometer
ISO 18118:2004	SCA – Auger electron spectroscopy (AES) and X-ray photoelectron spectroscopy (XPS) – guide to the use of experimentally determined relative sensitivity factors for the quantitative analysis of homogeneous materials
ISO/TR 18392:2005	SCA – XPS – procedures for determining backgrounds
ISO/TR 18394:2006	SCA – AES – derivation of chemical information
ISO 18516:2006	SCA – AES and XPS – determination of lateral resolution
ISO 19318:2004	SCA – XPS – reporting of methods used for charge control and charge correction
ISO/TR 19319:2003	SCA – AES and XPS – determination of lateral resolution, analysis area, and sample area viewed by the analyzer
ISO 20903:2006	SCA – AES and XPS – methods used to determine peak intensities and information required when reporting results
ISO 14237:2000	SCA – SIMS – determination of boron atomic concentration in silicon using uniformly doped materials
ISO/DIS 14237	SCA – SIMS – determination of boron atomic concentration in silicon using uniformly doped materials
ISO 17560:2002	SCA – SIMS – method for depth profiling of boron in silicon
ISO 18114:2003	SCA – SIMS – determination of relative sensitivity factors from ion-implanted reference materials
ISO 20341:2003	SCA – SIMS – method for estimating depth resolution parameters with multiple delta-layer reference materials
ISO 23812:2009	SCA – SIMS – method for depth calibration for silicon using multiple delta-layer reference materials

(continued)

Table 6.14 (continued)

Ref.	Title
ISO 23830:2008	SCA – SIMS – repeatability and constancy of the relative-intensity scale in static SIMS
ISO/DIS 10810	SCA – XPS – guidelines for analysis
ISO 15470:2004	SCA – XPS – description of selected instrumental performance parameters
ISO 15471:2004	SCA – AES – description of selected instrumental performance parameters
ISO 15472:2001	SCA – XPS – calibration of energy scales
ISO 15472:2001/DAmd 1	
ISO 17973:2002	SCA – medium-resolution AES – calibration of energy scales for elemental analysis
ISO 17974:2002	SCA – high-resolution AES – calibration of energy scales for elemental and chemical-state analysis
ISO 21270:2004	SCA – XPS and AES – linearity of intensity scale
ISO 24236:2005	SCA – AES – repeatability and constancy of intensity scale
ISO 24237:2005	SCA – XPS – repeatability and constancy of intensity scale
ISO 14707:2000	SCA – glow discharge optical emission spectroscopy (GD-OES) – introduction to use
ISO/TS 15338:2009	SCA – glow discharge mass spectrometry (GD-MS) – introduction to use
ISO 16962:2005	SCA – analysis of zinc- and/or aluminum-based metallic coatings by GD-OES

SCA surface chemical analysis

delta layers as a reference material to evaluate the extent of the above profile shift accurately, as shown in Fig. 6.11 [23].

Figure 6.12 shows schematically the relationship between the sputtered depth and sputtered time in the shallow region with the shift distance indicated by L_s when a sample with multiple delta layers is measured. The graph indicates that, in general, the average sputtering rate is larger at the near surface region, but reaches steady state value r_s after a few nm. When the average sputtering rate of the i-th layer, r_i, is regarded as r_s for the n-th layer and deeper, the sputtered depth z for the sputtering time t is given by $z = L_s + r_s t$. The calibration using a different sputtering rate to the reference specimen is also given. The estimation of peak shift due to atomic mixing as well as that due to peak coalescence and the derivation of uncertainty of the calibrated depth based on the Student's t-distribution are explained in detail in Annex B, C, and D of ISO 23812, respectively.

A potential candidate of the reference material is a film sample consisting of multi-layers of silicon (Si) and boron nitride (BN) delta layers on an appropriate substrate, e.g. silicon wafer. Typical thicknesses of the layers are designed at 8 nm for Si and less than 0.1 nm for BN, respectively. A prototype of the sample was fabricated by using the sputter deposition technique. The accurate evaluation of the thickness and density of each layer by using transmission electron microscope (TEM) images and the X-ray reflection method has been investigated to verify the

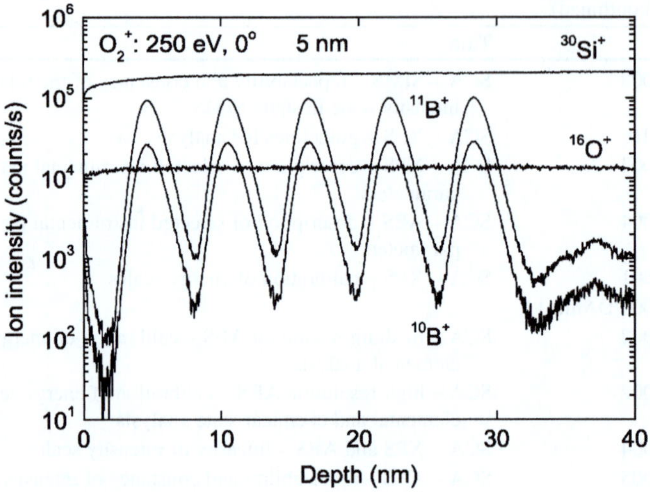

Fig. 6.11 SIMS depth profile of boron-delta layers in silicon [23]. © Crown copyright 2003. Reproduced by permission of the Controller of HMSO and the Queen's printer for Scotland

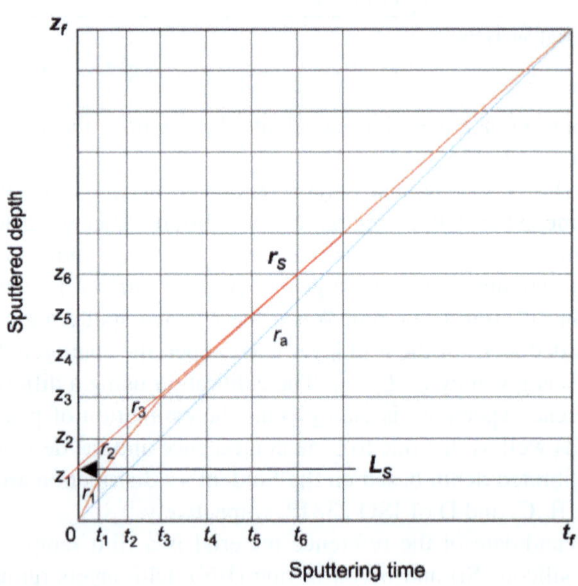

Fig. 6.12 Schematic drawing showing the relationship between the sputtered depth and sputtering time in the shallow region. The figure taken from ISO 23812:2009 is reproduced with the permission of the International Organization for Standardization, ISO. This standard can be obtained from ISO member (Japanese Standards Association: http://www.jsa.or.jp) and from the Web site of the ISO Central Secretariat at the following address: http://www.iso.org. Copyright remains with ISO

validity of the prototype. ISO 23812 was considered originally for the shallow depth profiling by SIMS, but since it provides an important measure for the nanometer depth profile it can be reasonably applied to the calibration of the measured size of nanostructured materials.

6.3.4 On-Going Project in ISO/TC 201 for Characterization of Nanostructured Materials

Finally, we introduce briefly, ISO TR, SCA – characterization of nanostructured materials, which is now under development by SC 5 (AES), designated as WD 14187. The introduction of the working draft states that because the large percent of nanostructured materials is associated with a surface or interface, the wide range of tools developed for surface characterization could be applied to these materials, but there have been two issues to overcome: (1) many of the tools of the necessary degree of spatial resolution in three dimensions needed to analyze individual nanostructured materials; and (2) the tools are sometimes applied to nanostructured materials without considering a range of analysis challenges or issues that these materials present. The TR is intended to give technical guidance to the issues and when published, it will address the types of information that surface analysis methods can provide about nanostructured materials as well as examine some of the technical challenges faced when applying surface analysis tools for characterization of nanostructured materials.

As mentioned earlier in this chapter, surfaces and interfaces can strongly influence many properties of materials and because of the importance of surfaces and interfaces a set of special tools has been developed to determine their compositions and to assess how these affect the properties of natural and engineered materials. Since nanostructured materials inherently involve a high percentage of surface or interface area, their properties are significantly influenced by the nature and properties of these surfaces and interfaces and therefore, the surface analysis techniques are said to be essential for revealing the properties of nanostructured materials. Among many kinds of surface-analysis techniques, the TR highlights AES, SIMS, SPM, and XPS. These techniques have different spatial resolutions and provide different types of information as summarized in Fig. 6.13 taken from the UK National Physical Laboratory website [24].

The TR then discusses the application of surface analysis capabilities to nanostructured materials by categorizing the nanostructures into nanofilms, layers or dispersions, and nanoparticles, and by listing information available for each structure from a specific surface analysis technique. They are summarized as follows:

1. Nanofilms, layers or dispersions
   Surface analysis tools considered in the TR were developed to enable the characterization of the outer few nanometers of materials and typically have depth resolutions of 1 nm or less. Consequently these techniques provide information about the composition, structure, chemical state and depth distribution of nanolayers on

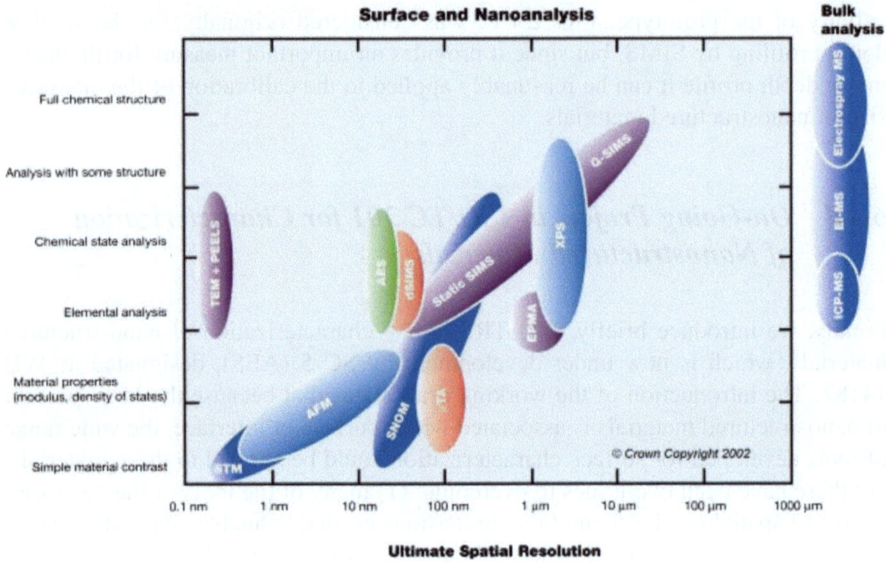

Fig. 6.13 Diagram providing over view of spatial resolution and types of information that can be obtained by a range to tools important for the analysis of nanostructured materials [24]. © Crown copyright 2010. Reproduced by permission of the Controller of HMSO and the Queen's printer for Scotland

surfaces, which is essential for characterizing the advanced materials used in microelectronics, etc. One successful example of the application of XPS to extract information about the nature of the elemental distribution with depth from a surface is a work by Tougaard. Although ion sputtering depth profiles can be used to extract some of this information more directly (see ISO/TR 22355:2007 in Table 6.13), Tougaard showed that XPS could be used to obtain quantitative information about the nanostructure of thin films based on inelastic scattering of the photo electrons, as schematically explained in Fig. 6.14 [25]. When combined with high-lateral resolution methods, the Tougaard background-analysis approach can be used to obtain a three-dimensional image of the composition of the near surface region of a complex film or material [26]. Other examples of applications of surface analysis to detect nanolayers or thin films are listed for specific materials with obtainable information by each technique as shown in Table 6.15.

2. Nanoparticles
   Although many variants of SPM methods are known to provide information from objects with nanometer dimensions, XPS and SIMS can also provide important information about nanoparticles, as listed below.

   (a) Information available from XPS of nanoparticles include the following:

      - Contamination, particle coatings and oxidation rates: If the particle shape is known, it is possible to obtain quantitative information about the thickness of a contamination layer or particle coating.

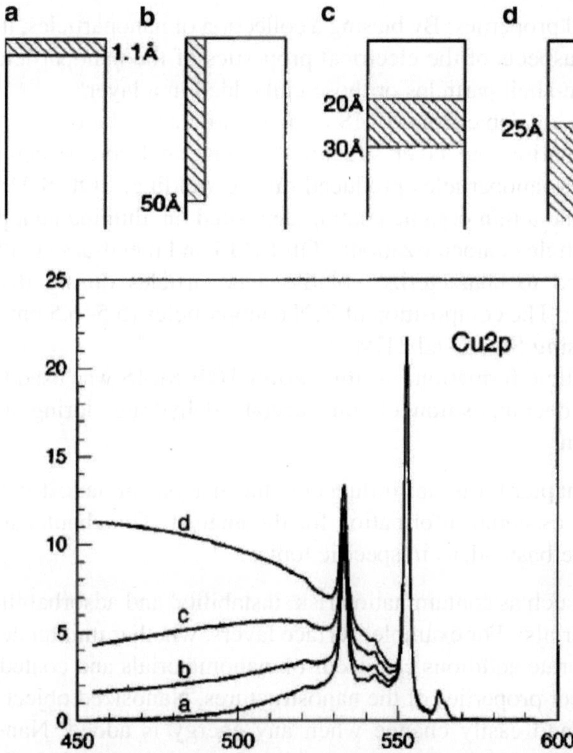

Fig. 6.14 Although these near surface elemental distribution of Cu in Au all produce the same Cu 2p photo-peak intensity, the differences in background below the peak provides information on the elemental distribution with depth [25]. Reprinted with permission from [25]. Copyright 1996, American Institute of Physics

Table 6.15 Examples of applications of surface analysis to characterization of nanofilms

System	Property	Technique
Metal alloy	Heating induced segregation	XPS
Langmuir Blodget film on glass	Thickness and structure	XPS
Nanocomposite for PEM fuel cell	Nanoparticle dispersion and composite aging	XPS
Self assembled monolayer	Functional group termination and layer structure	XPS
Self assembled monolayer	Domain structure	STM
Self assembled monolayer	Coverage, substrate interaction	XPS
SiO_2	Ultrathin layer thickness and uniformity	XPS
NiCr	Corrosion film properties	XPS
Immiscible polymers	Phase separation, annealing effects, domain structure, surface segregation, topography	XPS, AFM, TOF-SIMS

- Particle size: The ratio of photoelectron intensities form spherical particles having different escape depths can be used to approximate the particle size. The QUASES program can be available. It is also useful to know the particle shape which may be determined by TEM and possibly by SPM.

- Electrical properties: By biasing a collection of nanoparticles, it is possible to learn aspects of the electrical properties of the nanoparticles, particularly core-shell particles or those embedded in a layer.

(b) Information available from SIMS of nanoparticles include:
- Contamination and layer structure: Surface and core structure of large (300 nm) nanoparticles produced during welding. TOF-SIMS was used to examine a thin organic coating deposited on alumina nanoparticles.
- Nanoparticle characterization: TOF-SIMS and metal-assisted SIMS have been used to characterize volatile nanoparticles during diesel engine operation. The composition of ZrN nanoparticles (5.5–6.5 nm) was determined using SIMS and TEM.
- Nanoparticle formation: In situ thermo-TOF-SIMS was used to study the thermal decomposition of zinc acetate dehydrate during nanoparticle formation.

The TR has a chapter for issues influencing the analysis of nanostructured materials as useful and essential information for the analyst. The chapter has separate sections to describe basic ideas in specific topics:

1. General issues such as contamination risk, instability, and adsorbability of nanostructured materials: For example, surface layers, whether unintended contamination or deliberate additions can occur on nanomaterials and coated nanosized objects can affect properties of the nanostructures. Nanosized objects are inherently unstable and easily change when any energy is added. Nanostructured materials may adsorb solvents to a very high degree, altering their properties in various ways.
2. Importance of surface layers and surface chemistry which may be underemphasized compared with the novel properties of nanostructured materials themselves: Because of the high proportion of surface and interfaces associated with nanostructured materials, surfaces and interfaces can and do play an especially large role in the behaviors of their materials, such as nanotoxicity. However, because of the impacts of novel properties such as size-induced quantum states as well as obscure distinction between "bulk" and "surface" properties of the nanostructured materials, the importance of surface layers and surface chemistry is often more or less ignored.
3. Confluence of energy scales as many types of energy including thermal, chemical, mechanical, magnetic, and electrostatic energies plotted in a common scale may converge for sizes of nanostructured materials (see Fig. 6.15): For objects of sizes associated with nanotechnology, many of the energy scales converge providing many opportunities for coupling of different modes of excitation. Therefore, there is a significant probability that probe effect, environmental effects, or near-neighbor effects can influence the properties of the nano-sized objects.
4. Influence of shape especially in the XPS measurement of nanoparticles where, for example, a uniform coating on spherical particles would produce a different ratio of surface to substrate signals than a flat plate: When the particle sizes are

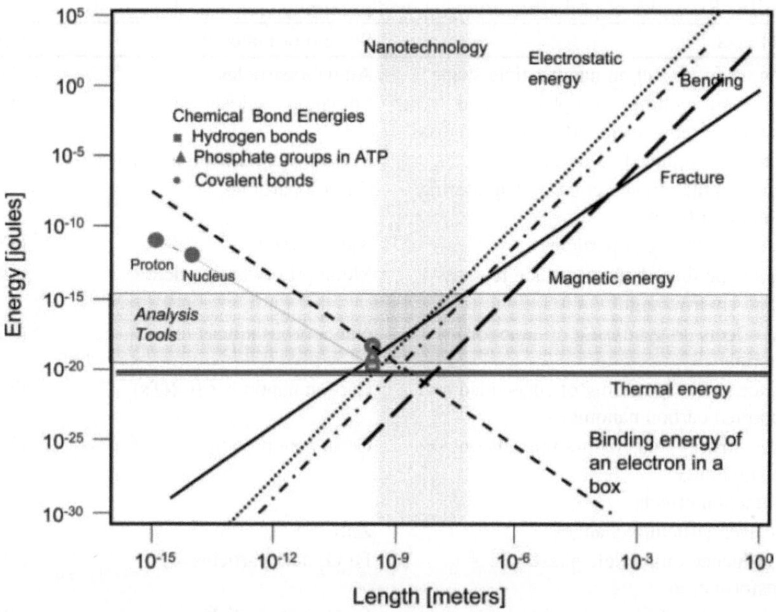

Fig. 6.15 Variation in thermal, chemical, mechanical, magnetic, and electrostatic energies as a function of the size of an object [27]. © Crown copyright 2003. Reproduced by permission of the Controller of HMSO and the Queen's printer for Scotland

sufficiently small, some photoelectrons can travel through the whole particles to give stronger signals. For nanoparticles larger than the electron inelastic mean free path, XPS analysis of a collection of nanoparticles might be considered or approximated as a characterization of a rough surface.

5. Particle stability in terms of shape, crystal structure, and damage by an electron beam: The energy of a nanoparticle shows many local-minima configurations corresponding to different structures and a small excitation may be sufficient to induce transitions of the particle. The observed crystal structures of nanoparticles may also be easily altered, even when constrained within a matrix. Sometimes clearly irreversible changes are found in the material being examined and these changes usually involve loss of information about the initial particles and, from the analysis perspective, must be considered as probe damage.
6. Effect of environment on nanomaterials structure and properties, time dependent properties, and proximity effects with substrate, buffered layer, etc.: There is a growing list of experimental observations of environmentally induced changes in the physical and chemical properties of nanostructured materials systems such as the structure change of ZnS nanoparticles in wet and dry environments, moisture-induced decreasing size for which phase transitions are observed for Fe_2O_3 nanoparticles, etc. The properties of individual nano-sized objects can be significantly altered when they are supported on a substrate, collected into aggregates, or

Table 6.16 Examples of probe, environment, and proximity effects [27]

Probe effects	System or material	Ref.
Electron beam impact on nanoparticle shape	Au nanoparticles	[28, 29]
Electron beam melting, amorphizing, and crystallization of nanoparticles in a matrix	Sn nanoparticles in SiO_2	[30]
Electron beam-induced oxidation	FeO/FeO_x core/shell nanoparticles	[31]
Ion beam interaction and enhanced sputtering of small particles	Carbon particles	[32]
Enhanced sputtering of particles	NaCl crystals	[33]
Sputter sharpening of steep surface features	Metal pit or "antiparticle"	[34]
Probe and environment		
Solvent effects on sputtering of nanoporous materials	Nanoporous silica	[35]
Difference in the sputtering of suspended and supported carbon nanotubes	Carbon nanotubes (CNTs)	[36]
Specimen history and coating impacts on X-ray damage	Ceria nanoparticles	[37]
Environmental effects		
Water driven structure changes	ZnS	[38]
Water influence on particle phase transformation	Fe_2O_3 nanoparticles	[39]
Nanotube encapsulation impact on iron oxide reduction temperature	Fe_2O_3 nanoparticles	[40]
Humidity effects on polymer nanostructures	Poly(vinyl butyral) (PVB) and poly(methylmethacrylate) (PMMA) mixtures	[41]
Surface sorbate effects on growth shape	Solution grown nanoparticles	[42]
Surface sorbate effects on particle separation	Oxide and metal nanoparticles	[43]
Environmental impact on particle chemical state	Ceria nanoparticles	[44, 45]
Proximity or distance effects		
Charge buildup or accumulation during XPS	Nanoparticles on insulating substrates and at interfaces	[46, 47]
Plasmon coupling – basis of a nanoruler	Au nanoparticles	[48, 49]
Coupling and engaging of quantum states	Quantum dot molecules	[50]
Impact of spacing and aggregation on magnetic properties	Iron oxide nanoparticles	[43, 51]
Interphase effects on composite properties	Nanoparticle dispersion in composites	[52]
Effect of "buffer layers" on the optical properties of silicon nanocrystal superlattices	Si-rich oxide and SiO_2	[53]

possibly assembled into a composite and these are known as proximity effects. A list with more items is given in a separate literature as shown in Table 6.16 [27].

The TR when published will provide useful and essential guideline for characterization of nanostructured materials using surface chemical analysis tools.

6.4 Application Measurement in Other Standard Organizations

6.4.1 *International Workshop on Documentary Standards for Measurement and Characterization*

Measurement and characterization of nanotechnology relates to not only the standard organizations mentioned above, but also other standard organizations such as ISO/TC 24/SC 4 on sizing by methods other than sieving, for example, and ISO/TC 202 on Microbeam analysis. ISO 12025 of ISO/TC 229 relates closely to the activity of TC 24/SC 4, and TS 13126 (artificial gratings used in nanotechnology: description and measurement of dimensional quality parameters) of ISO/TC 229 and IEC/TC 113 relates to the activity of ISO/TC 201/SC 9 on scanning probe microscopy [8]. ISO/TC 229/WG 2 TS 10797 and TS 10798, the projects on SWCNT, have a close relationship to TC 202 on microbeam analysis.

In order to promote collaboration among nanotechnology standards organizations, the International Workshop on Documentary Standards for Measurement and Characterization for Nanotechnologies was held at NIST/USA in February of 2008 under the co-sponsorship of ISO, IEC, NIST, and OECD. Through discussion at this workshop, the following were identified as important for enhancing the development, efficacy, harmonization and uptake of documentary standards relevant to the field of measurement and characterization for nanotechnologies:

1. Communication and coordination (within and between the various standards development organizations and with interested metrology institutes)
2. Repository of information (on existing standards and standardization projects)
3. Development of a terminology and definitions database (freely accessible and searchable)
4. Participation of stakeholders (in identifying and verifying standards needs)
5. Consideration of instruments (in investigating the implications for human health and environmental safety of manufactured nanomaterials)

As to communication and coordination, IEEE agreed to develop a discussion forum that would be continually updated, to align information and developments from the different standards developing organizations. ISO agreed to develop a platform for managing terminology(ies) for nanotechnologies, through the new "ISO Concept Database". Moreover, it was also noticed that good practice/guidance documents are important as well as standard documents that cover:

1. Information for handling/using, stability, and concentration (together with it's definition) of nanoparticles
2. Suites of measurement techniques (and the information the combined data set might provide)
3. Sample preparation (in consideration of dispersion and aggregation/agglomeration and for human and eco-toxicology testing)

4. Application and limitation of surface analysis to nanoparticles
5. Dose measurement and dosimetry for in vitro and in vivo human and eco-toxicology studies

In items (2) and (5), which relate to specific areas of human health and toxicology, there is a need for greater dissemination, verification and validation of handling and testing protocols and related procedures by the broader community.

It was proposed that ISO/TC 229 establish a Nanotechnologies Liaison Coordination Group (NLCG), which would coordinate and harmonize the work of relevant TCs in the field of nanotechnologies, and to identify cross-cutting gaps and opportunities and ways to address these. Held in conjunction each ISO/TC 229 plenary meeting, discussions on related projects among these TCs have benefited from the collection of comments and opinions from TC members.

6.4.2 Activities of ISO/TC 24/SC 4 (Standardization of Particle Characterization)

Since its establishment in 1947, ISO/TC 24 (particle characterization including sieving) has long been working on standardization of equipments and methods used in size classification of particulate materials. At its early stage, equipments and methods of orthodox concept based on sieving were the targets of standardization, but a growing demand to handle particles of smaller sizes (due to nanotechnology) has caused TC 24 to develop standards for wider variety of equipments and methods. The business plan of TC 24 revised in 2004 clearly mentions the recent issues on particle characterization and a need to develop new types of standards to solve them as follows:

The scope covers standardization pertaining to equipment and methods used in size classification of particulate material in solid or liquid state. Particle size analysis and characterization is intensely used in almost all industrial processes and productions (e.g. production of cement) or other processed material which is grinded, milled or crushed. The chemical process industries alone include large multi-national corporations. Some 80% of their products, employees, and international trade rely on an accurate knowledge of particle size distribution for success. Industry environmental agencies, hospitals, and universities all need good procedures for dispersing powders and stabilizing the resulting suspensions in liquid if they are to obtain the accurate particle size distributions vital to fulfilling their production, application, or research functions.

In the past 10 years the technology of particle size measurement has changed considerably as follows:

1. Many particulate products with new chemical compositions have been introduced (catalysts, reinforcing fibers, superconductors)
2. Many major products have been introduced using particles of smaller sizes (ceramics, electronics, photography, nanoparticles)

3. Product specifications have become more restrictive, so that more accurate particle size analyses are now required
4. The introduction of inexpensive computers has allowed both novel and more sophisticated methods for measuring particle size distribution
5. New classes of chemicals have been introduced for dispersing powders in liquids (star polymers, dispersants based on group transfer polymerization, geminal multifunctional dispersants). For example, the availability of thousands of possible dispersing agents, and many techniques for deagglomerating powders in liquids has made it very difficult for analysts faced with a new powder to decide which deagglomeration method and which stabilizing surfactant are likely to be successful in making a stable dispersion in the liquid used with a specific particle size analysis method. Moreover, standards on tests for particle size analysis and characterization are required for quality assurance, accreditation and certification"

TC 24 has now one Working Group for Terminology and two Subcommittees, SC 4 for particle characterization and SC 8 for test sieves, sieving and industrial screens. Within SC 4, 17 Working Groups have been established and most of them, i.e., 15 WGs, are still active as listed in Table 6.17.

As summarized in Tables 6.18 and 6.19, ISO/TC 24/SC 4 have published 34 International Standards (ISs) and five projects are under development. From this list, it can be seen that the work area of standardization by TC 24/SC 4 has expanded from simple sieving (the related standards are not shown as they are no longer active) to various methods to meet the needs from forefront industries using new materials.

Table 6.17 Working groups in ISO/TC 24/SC 4

Subcommittee/working group	Title
TC 24/SC 4/WG 1	Representation of analysis data
TC 24/SC 4/WG 2	Sedimentation, classification
TC 24/SC 4/WG 3	Pore size distribution, porosity
TC 24/SC 4/WG 5	Electrical sensing methods
TC 24/SC 4/WG 6	Laser diffraction methods
TC 24/SC 4/WG 7	Dynamic light scattering
TC 24/SC 4/WG 8	Image analysis methods
TC 24/SC 4/WG 9	Single particle light interaction methods
TC 24/SC 4/WG 10	Small angle X-ray scattering method
TC 24/SC 4/WG 11	Sample preparation and reference materials
TC 24/SC 4/WG 12	Electrical mobility and number concentration analysis for aerosol particle
TC 24/SC 4/WG 14	Acoustic methods
TC 24/SC 4/WG 15	Particle characterization by focused beam techniques
TC 24/SC 4/WG 16	Characterisation of particle dispersion in liquids
TC 24/SC 4/WG 17	Methods for zeta potential determination

Table 6.18 ISs published by TC 24/SC 4

Published ISs	Titles
ISO 9276-1:1998	Representation of results of particle size analysis – part 1: graphical representation
ISO 9276-1:1998/ Cor 1:2004	
ISO 9276-2:2001	Representation of results of particle size analysis – part 2: calculation of average particle sizes/diameters and moments from particle size distributions
ISO 9276-3:2008	Representation of results of particle size analysis – part 3: adjustment of an experimental curve to a reference model
ISO 9276-4:2001	Representation of results of particle size analysis – part 4: characterization of a classification process
ISO 9276-5:2005	Representation of results of particle size analysis – part 5: methods of calculation relating to particle size analyses using logarithmic normal probability distribution
ISO 9276-6:2008	Representation of results of particle size analysis – part 6: descriptive and quantitative representation of particle shape and morphology
ISO 9277:1995	Determination of the specific surface area of solids by gas adsorption using the BET method
ISO 13317-1:2001	Determination of particle size distribution by gravitational liquid sedimentation methods – part 1: general principles and guidelines
ISO 13317-2:2001	Determination of particle size distribution by gravitational liquid sedimentation methods – part 2: fixed pipette method
ISO 13317-3:2001	Determination of particle size distribution by gravitational liquid sedimentation methods – part 3: X-ray gravitational technique
ISO 13318-1:2001	Determination of particle size distribution by centrifugal liquid sedimentation methods – part 1: general principles and guidelines
ISO 13318-2:2007	Determination of particle size distribution by centrifugal liquid sedimentation methods – part 2: photocentrifuge method
ISO 13318-3:2004	Determination of particle size distribution by centrifugal liquid sedimentation methods – part 3: centrifugal X-ray method
ISO 13319:2007	Determination of particle size distributions – electrical sensing zone method
ISO 13320:2009	Particle size analysis – laser diffraction methods
ISO 13321:1996	Particle size analysis – photon correlation spectroscopy
ISO 13322-1:2004	Particle size analysis – image analysis methods – part 1: static image analysis methods
ISO 13322-2:2006	Particle size analysis – image analysis methods – part 2: dynamic image analysis methods
ISO/TS 13762:2001	Particle size analysis – small angle X-ray scattering method
ISO 14488:2007	Particulate materials – sampling and sample splitting for the determination of particulate properties
ISO 14887:2000	Sample preparation – dispersing procedures for powders in liquids
ISO 15900:2009	Determination of particle size distribution – differential electrical mobility analysis for aerosol particles
ISO 15901-1:2005	Pore size distribution and porosity of solid materials by mercury porosimetry and gas adsorption – part 1: mercury porosimetry
ISO 15901-1:2005/ Cor 1:2007	

(continued)

Table 6.18 (continued)

Published ISs	Titles
ISO 15901-2:2006	Pore size distribution and porosity of solid materials by mercury porosimetry and gas adsorption – part 2: analysis of mesopores and macropores by gas adsorption
ISO 15901-2:2006/ Cor 1:2007	
ISO 15901-3:2007	Pore size distribution and porosity of solid materials by mercury porosimetry and gas adsorption – part 3: analysis of micropores by gas adsorption
I SO 20998-1:2006	Measurement and characterization of particles by acoustic methods – part 1: concepts and procedures in ultrasonic attenuation spectroscopy
ISO 21501-1:2009	Determination of particle size distribution – single particle light interaction methods – part 1: light scattering aerosol spectrometer
ISO 21501-2:2007	Determination of particle size distribution – single particle light interaction methods – part 2: light scattering liquid-borne particle counter
ISO 21501-3:2007	Determination of particle size distribution – single particle light interaction methods – part 3: light extinction liquid-borne particle counter
ISO 21501-4:2007	Determination of particle size distribution – single particle light interaction methods – part 4: light scattering airborne particle counter for clean spaces
ISO 22412:2008	Particle size analysis – dynamic light scattering (DLS)

Table 6.19 ISs under development by TC 24/SC 4

ISs under development	Titles
ISO/FDIS 9277	Determination of the specific surface area of solids by gas adsorption – BET method
ISO/CD 13099-1	Methods for zeta potential determination – part 1: introduction
ISO/CD 13099-2	Methods for zeta potential determination – part 2: optical method
ISO/NP 13322-1	Particle size analysis – image analysis methods – part 1: static image analysis methods
ISO/CD 26824	Particle characterization of particulate systems – vocabulary

6.4.3 IEC/TC 113

IEC TC 113 currently has three WGs, two of which have started joint activities with ISO/TC229/WG 1 and WG 2 as JWG 1 and JWG 2, and one additional WG 3 for performance assessment. TC 113/WG 3 has in its work program, two joint activities with the IEEE. Published and current work items are summarized in Table 6.20. Among them, IEC 62624 is a joint publication with IEEE, and JPT 62607, JPT 10797, JPT 62622, and JPT 13278 are documents prepared jointly between ISO/TC 229 and IEC/TC 113. The project for "artificial gratings used in nanotechnology: description and measurement of dimensional quality parameters" is led by Dr. H. Bosse/Germany, and "nanomanufacturing – key control characteristics

Table 6.20 Published and current working items of IEC TC 113

Project	Title	Status	Remark
IEC 62624	Test methods for measurement of electrical properties of carbon nanotubes	P	[a]
IEC/TR 113-69	Nanoscale electrical contacts	Preliminary	
IEC/TR 113-70	IEC nano-electronics standards roadmap	Preliminary	
IEC/TR 62565-1	Nanomanufacturing – material specifications – part 1: basic concept	Preliminary	
IEC/PAS 62565-2-1	Nanomanufacturing – material specifications – part 2-1: single wall carbon nanotubes – blank detail specification	PAS	
IEC/TS 62607-2-1	Nanomanufacturing – key control characteristics – part 2-1: carbon nanotube materials – film resistance	WD	ISO/IEC TS 62607[b]
IEC/TS 113-82	Nanomanufacturing – key control characteristics of luminescent nanomaterials part 3-1: quantum efficiency	Preliminary	
IEC/TS 62622	Artificial gratings used in nanotechnology: description and measurement of dimensional quality parameters	WD	ISO/IEC TS 13126[b]
IEC/TS 62659	Proposal of large scale manufacturing of nanoelectronics	WD	
ISO/TS 13278	Carbon nanotubes – determination of metal impurities in carbon nanotubes (CNTs) using inductively coupled plasma-mass spectroscopy (ICP-MS)	WD	[b]
ISO/TS 10797	Nanotubes – use of transmission electron microscopy (TEM) in the characterization of single walled carbon nanotubes (SWCNTs)	WD	[b]

P published document, *WD* working draft under preparation, *Preliminary* new work item proposal under preparation

[a]Joint IEC-IEEE project or published standard

[b]Joint IEC/TC 113 and ISO/TC 229 projects

for carbon nanotube materials – film resistance" is led by Dr. H. Jin Lee/Korea (special project between ISO/TC 229 and IEC/TC 113).

PT 62565-2-2, PT 62607-3-1 and JPT 62659-2 are projects prepared exclusively within IEC/TC 113. These are the first of what is expected to be many standards and technical specifications to be developed to facilitate the mass manufacture of nanotechnology enabled electrical and electronic end products and subassemblies.

PT 113-70 is an IEC technical report under development, describing the state of the art in the field of nanoscale electrical contacts, and the critical nature of these contacts in facilitating interaction between nanoscale and macroscale subassemblies. The project is being led by Dr. G. Monty, IEC/TC 113 Chairman.

Table 6.21 Current projects of CEN-352 (as of 16 October 2009)

Project no	Project title	Lead
CEN/ISO TR 11808	Guide to nanoparticle measurement methods and their limitations	CEN
CEN/ISO TR 11811	Guide to methods for nanotribology measurements	CEN
CEN/ISO TR 13830	Guidance on labelling of manufactured nanoparticles and products containing manufactured nanoparticles	CEN

6.4.4 CEN/TC 352

As is briefly introduced in the Sect. 6.1, CEN has established the new TC 352 on Nanotechnology in 2005. It focuses on to develop a set of standards addressing the following aspects of nanotechnologies:

1. Classification, terminology and nomenclature
2. Metrology and instrumentation, including specifications for reference materials
3. Test methodologies
4. Modeling and simulation
5. Science-based health, safety and environmental practices
6. Nanotechnology products and processes

It has to be mentioned that topics of mutual interest to ISO/TC 229 and CEN/TC 352 are expected to be carried out under the Vienna Agreement with CEN or ISO lead. Table 6.21 shows the list of current projects which has been approved as work items by ISO/TC 229 and are under development by CEN/TC 352 ledership. TR 11808 and TR 11811 relate to the activity of ISO/TC 229/JWG 2, while TR 13830 relates to that of ISO/TC 229/WG 4.

6.4.5 ASTM International E42 and E56 Committees

Similar to the ISO structure, the ASTM International has over 130 technical committees to establish voluntarily consensus technical standards for safer, better and more cost-effective products and services. Among the ASTM technical committees, E42 (surface analysis) and E56 (nanotechnology) conduct standardization activities that can be compared to those of ISO/TC 201 and TC 229. ASTM/E42 and its 12 subcommittees cover a greater part of the ISO/TC 201 and TC 202 activities as shown in Table 6.22, while E56, formed in 2005, has a structure similar to that of ISO/TC 229, as shown in Table 6.23.

ASTM/E42.14 (STM/AFM) which can be regarded as a counterpart of ISO/TC 201/SC 9 (SPM) already has three active standards, as follows;

1. E1813-96 (2007) "Standard practice for measuring and reporting probe tip shape in scanning probe microscopy"
2. E2382-04 "Guide to scanner and tip related artifacts in scanning tunneling microscopy and atomic force microscopy"

Table 6.22 Subcommittees of the ASTM E42 committee

SC	Title	Number of active standards
E42.02	Terminology	1
E42.03	Auger electron spectroscopy and X-ray photoelectron spectroscopy	13
E42.06	SIMS	9
E42.08	Ion beam sputtering	3
E42.13	Vacuum technology	0
E42.14	STM/AFM	3
E42.15	Electron probe microanalysis/electron microscopy	0
E42.90	Executive	0
E42.91	Awards	0
E42.92	US TAG ISO/TC 201	0
E42.94	US TAG ISO/TC 112	0
E42.96	US TAG ISO/TC 202	0

Table 6.23 Subcommittees of the ASTM E56 committee

SC	Title	Number of active standards
E56.01	Informatics and terminology	1
E56.02	Characterization: physical, chemical, and toxicological properties	5
E56.03	Environment, health, and safety	1
E56.04	International law and intellectual property	0
E56.05	Liaison and international cooperation	0
E56.90	Executive	0
E56.91	Strategic planning and review	0

3. E2530-06 "Standard practice for calibrating the *z*-magnification of an atomic force microscope at subnanometer displacement levels using Si(111) monatomic steps"

These standards may belong to a basic standard system for STM and AFM, which ISO/TC 201/SC 9 also aims to establish. E1813-96 in particular can be compared with AWI 13095 "procedure for in situ characterization of AFM probes used for nanostructure measurement," described in the previous section for the similarity in the scopes as shown in Table 6.24.

On the other hand, ASTM/E56.02 (characterization: physical, chemical, and toxicological properties) which can be regarded as a counterpart of ISO/TC229/WG2 has already published following five documents:

1. E2490-09 "Standard Guide for Measurement of Particle Size Distribution of Nanomaterials in Suspension by Photon Correlation Spectroscopy (PCS)"
2. E2524-08 "Standard Test Method for Analysis of Hemolytic Properties of Nanoparticles"
3. E2525-08 "Standard Test Method for Evaluation of the Effect of Nanoparticulate Materials on the Formation of Mouse Granulocyte-Macrophage Colonies"

Table 6.24 Scope of ASTM/E1813-96 and ISO/AWI 13095

ASTM/E1813-96	ISO/AWI 13095
1.1 This practice covers scanning probe microscopy and describes the parameters needed for probe shape and orientation 1.2 This practice also describes a method for measuring the shape and size of a probe tip to be used in scanning probe microscopy. The method employs special sample shapes, known as probe characterizers, which can be scanned with a probe microscope to determine the dimensions of the probe. Mathematical techniques to extract the probe shape from the scans of the characterizers have been published (2–5) This standard does not purport to address all of the safety concerns, if any, associated with its use. It is the responsibility of the user of this standard to establish appropriate safety and health practices and determine the applicability of regulatory limitations prior to use	This international standard specifies a method for characterizing the shape of AFM probes. This is important for measuring the shapes of three-dimensional nano-structures. This characterizing method is related to the use of a reference material with a comb-shape structure and an isolated narrow ridge structure. The method provides the cross-sectional profile of the AFM probe in a given direction by determining the relation between the probe width (W1, W2) and the probe length (L1, L2) measured from the probe apex. The method is appropriate to characterize the profile of a probe with widths between several nm and a few hundred nm. The method is intended to reduce the uncertainty of AFM measurement of nano-materials or nano-structures

4. E2526-08 "Standard Test Method for Evaluation of Cytotoxicity of Nanoparticulate Materials in Porcine Kidney Cells and Human Hepatocarcinoma Cells"
5. E2578-07 "Standard Practice for Calculation of Mean Sizes/Diameters and Standard Deviations of Particle Size Distributions"

Moreover, it is currently preparing additional three documents for "New Measurement of particle size distribution of nanomaterials in suspension by Photon Correlation Spectroscopy (PCS)" (WK8705), "New Guide for Zeta potential measurement by electrophoretic mobility" (WK21915), and for "New Guide for Measurement of particle size distribution of nanomaterials in suspension by nanoparticle tracking analysis (NTA)" (WK26321). It should be noted that three among five published documents relate to test method of nanoparticles in biological fields reflecting the title of E56.02.

6.4.6 IEEE Nanotechnology Standards Working Group

The Institute of Electrical and Electronics Engineers (IEEE) established the IEEE Standards Association (SA) to provide a standards program that serves the global needs of industry, government, and the public. IEEE-SA also works to assure the effectiveness and high visibility of this standards program both within the IEEE and throughout the global community.

Many working groups conduct standards projects in various areas within the scope of IEEE. Among them, the Nanotechnology Standards Working Group is developing nanotechnology-based electronics standards. Key drivers of the working group are the need for reproducibility of results, international collaboration, and common means for communicating across traditional scientific disciplines. This activity is part of a broader nanotechnology effort at the IEEE driven by the IEEE Nanotechnology Council (NTC), an interdisciplinary group whose members are drawn from 19 IEEE Societies. Two Study Groups (SG) operate under the Nanotechnology Standards Working Group.

The Materials Nanometrology SG considers all measurement areas needed to evaluate nanomaterials with standardized characterization and reporting methods for electrical properties, size and structure, thermal properties, composition, and surface properties.

The Nanoscale Devices SG considers:

1. Device Measurement including instrumentation, destructive and non-destructive testing, chemical and biological issues, quantum and contact effects, and mechanical, optoelectronic, electrical and thermal properties
2. Device Geometry for two terminal (e.g. diodes, LEDs, capacitors, actuators, resistors) and three terminal devices (e.g. transistors, memory cells and quantum cellular automata)

In addition, an interoperability SG will be formed to consider various interoperability environments (e.g. electrical, photonic, and mechanical) and to interface between elements of nanoscale devices and systems incorporating nanoscale devices. So far, one standard has been published and one is under development by the nanotechnology standards WG, as described below.

1. Published Standard: IEEE Standard 1650™-2005 "IEEE Standard Test Methods for Measurement of Electrical Properties of Carbon Nanotubes"
   Scope: This standard describes methods for the electrical characterization of carbon nanotubes. The methods are independent of processing routes used to fabricate the carbon nanotubes.
   Purpose: The purpose of this standard is to provide methods for the electrical characterization of carbon nanotubes and the means of reporting performance and other data. This is intended to provide and suggest procedures for characterization and reporting of data. These methods enable the creation of a suggested reporting standard that are used by research through manufacturing as the technology is developed. Moreover, the standards recommend the necessary tools and procedures for validation.
2. Standard Under Development: "Standard Methods for the Characterization of Carbon Nanotubes Used as Additives in Bulk Materials" (P1690TM)
   Scope: This project will develop standard methods for the characterization of carbon nanotubes used as additives in bulk materials. The methods will be independent of processing routes used to fabricate the carbon nanotubes.
   Purpose: The purpose of the proposed project is to provide and suggest procedures for characterization and reporting of data. These methods will enable the creation of a suggested reporting standard that will be used by research through

manufacturing as the technology is developed. Moreover, the standards will recommend the necessary tools and procedures for validation.

In 2007, the IEEE-SA established a formal liaison with IEC TC 113 WG3. Through this liaison, IEEE members enjoy the benefit of "expert" membership on IEC TC 113 WG3 project and maintenance teams established for standards and specifications developed jointly between the IEC and IEEE. Such is the case for the maintenance team for IEC/IEEE 62624 (IEEE Standard 1650™).

6.5 Conclusion

6.5.1 Standardization for Characterization from Nano-materials to Nano-intermediates

Based on the prediction for growth of the global nanotech market, a much higher increase in growth is expected for nano-intermediates (intermediate products with nanoscale features) than nanomaterials (nanoscale structures in unprocessed form) and for nano-application products (finished goods incorporating nanotechnology) in the technology value chain. Here, nanoparticles, nanotubes, quantum dots, fullerenes, dendrimers, and nanoporous materials, are considered as representatives of nanomaterials, while coatings, fabrics, memory and logic chips, contrast media, optical components, orthopedic materials, and superconducting wires, are considered as representative of the nano-intermediate [54]. The prediction of the increasing growth rate for nanomaterials and nano application products is already running as high as 25% per year. Given that the production of nano-intermediates from nanomaterials would contribute to the value-added process, it is critical to consider placing greater emphasis on standardization for the characterization of nano-intermediates.

The basic type of product architecture can be classified into the matrix shown in Table 6.25 [55]. Here, "modular architecture" is possible if one-to-one correspondence between functional and structural elements is achieved. This fits well to standardization either across firms (as open-modular type) or within a firm

Table 6.25 Basic type of product architecture

	Integral	Modular
Closed	Small Cars Motorcycle Game software Compact consumer electronics	Mainframe computer Machine tools LEGO (building block toy)
Open	New target (production of nano-intermediate from nanomaterials)	Personal computer Bicycle PC software Internet

Modified and Reprinted with permission from [55]

Table 6.26 URL's for the organizations mentioned in this chapter

CEN TC 352; http://www.cen.eu/cenorm/sectors/sectors/nanotechnologies/nanotechnologies.asp
CEN BT WG 166; http://www.cen.eu/cenorm/sectors/sectors/materials/nanotechnology.asp
ANSI NSP; http://www.ansi.org/standards_activities/standards_boards_panels/nsp/overview.aspx?menuid=3
ETUC (European Trade Union Confederation); http://www.etuc.org/a/5159?var_recherche=Nanotechnology
ETC group press release (July 31, 2007) "Broad International coalition issues urgent call for strong oversight of nanotechnology"; http://www.etcgroup.org/en/node/651
ASTM E42; http://www.astm.org/COMMIT/COMMITTEE/E42.htm
ASTM E56; http://www.astm.org/COMMIT/COMMITTEE/E56.htm
IEEE-SA; http://www.ieee.org/web/standards/home/index.html
IEEE nanotechnology standards; http://www.grouper.ieee.org/groups/nano/
ITRS; http://www.itrs.net/
NanoRoadSME; http://www.nanoroad.net/

(as closed-modular type). The production of personal computers and bicycles can be considered as typical examples of open-modular type architecture, while the production of mainframe computers and machine tools are considered examples of closed-modular type architecture. On the other hand "integral architecture" has to consider a one-to-many relationship between function and structural elements, and standardization activity for this setting is not currently in demand. The production of small cars, motorcycles, game software etc., is considered to be closed-integral type architecture.

Since production of nano-intermediates from nanomaterials needs in some case precise tuning of structure elements to serve a requested function, it might be classified into an integral type. It would be much more meaningful to consider the possibility of an open-integral type of architecture for the production on nano-intermediates, this presents the opportunity for nanomaterials to be provided in the open market. Standardization activities that have mainly been focused on the characterization of nano-materials would help the establishment of "open-integral type" production using well-characterized nanomaterials from many separate suppliers, leading to the growth of the nanotechnology industry. Activities relating nanotechnology including those of ISO/TC 229 is, therefore, highly anticipated (Table 6.26).

References

Section 1

1. Roco, M.: Nanotechnology R&D in the Americas and the global context. In: 2nd International Dialogue on Responsible Research and Development of Nanotechnology, Tokyo, Japan, 27–28 June 2006
2. Proffitt, F.: Yellow light for nanotech, Science **305**, 762 (2004)
3. Maynard, A.D.: Safe handling of nanotechnology, Nature **444**, 267–269 (2006)
4. ETUC: http://www.etuc.org/a/5159?var_recherche=Nanotechnology

5. Manna, S.K., Sarkar, S., Barr, J., Wise, K., Barrera, E.V., Jejelowo, O., Rice-Ficht, A.C., Ramesh, G.T.: Single-walled carbon nanotube induces oxidative stress and activates nuclear transcription factor-κB in human keratinocytes, Nano Letters **5**, 1676–1684 (2005)
6. Takagi, A., Hirose, A., Nishimura, T., Fukumori, N., Ogata, A., Ohashi, N., Kitajima, S., Kanno, J.: Induction of mesothelioma in p53+/- mouse by intraperitoneal application of multi-wall carbon nanotube, J. Toxicol. Sci. **33**, 105–116 (2008)
7. Poland, C.A., Duffin, R., Kinolch, I., Maynard, A., Wallace, W.A.H., Seaton, A., Brown, V.S., MacNee, W., Donaldson, K.: Carbon nanotubes introduced into the abdominal cavity of mice show asbestos-like pathogenicity in a pilot study, Nature Nanotechnology **3**, 423–428 (2008)
8. Hench, L.L.: In: Hench, L.L., Wilson, J. (eds.) An Introduction to Bioceramics, p. 319. World Scientific, Singapore (1993). Chapter 18: Characterization of Bioceramics
9. Hossain, K.: WG2 study group on strategy (2008), outline strategy for ISO TC 229 WG2 – nanotechnologies, ver. 8.0 (2008, unpublished)

Section 2

10. Committee to discuss protective actions for exposure to workers of chemical materials of which hazardous property to human body is not clearly identified, (Ministry of Health, Labor and Welfare, Nov. 26, 2011) Part 2 (in Japanese) http://www.mhlw.go.jp/shingi/2008/11/dl/s1126-6a.pdf (2009)
11. Iijima, S.: Helical microtubules of graphitic carbon, Nature **354**, 56–58 (1991)
12. Collins, P.G., Avouric, Ph.: Nanotubes for electronics, Sci, Am. **283**, 62–69 (2000)
13. Terrones, M.: Science and technology of the twenty-first century: Synthesis, Properties, and Applications of Carbon Nanotubes, Ann. Rev. Mater. Res. **33**, 419–501 (2003)
14. Osawa, E.: Kagaku (in Japanese) **25**, 854–863 (1970)
15. Kroto, H.W., Heath, J.R., O'Brien, S.C., Curl, R.F., Smalley, R.E.: C60 Buckminsterfullerene, Nature **318**, 162–163 (1985)
16. Chai, Y., Guo, T., Jin, C., Haufler, R.E., Chibante, L.P.F., Fure, J., Wang, L., Alford, J.M., Smalley, E.: Fullerenes with metals inside, J. Phys. Chem. **95**, 7564–7568 (1991)
17. Hinokuma, K., Ata, M.: Fullerene proton conductors, Chem. Phys. Lett. **341**, 442–446 (2001)
18. Hinokuma, K., Ata, M.: Proton conduction in polyhydroxy hydrogensulfated fullerenes, J. Electrochem. Soc. **150**, A112–A116 (2003)
19. Chikamatsu, M., Nagamatsu, S., Yoshida, Y., Saito, K., Yase, K., Kikuchi, K.: Solution-processed n-type organic thin-film transistors with high field-effect mobility, Appl. Phys. Lett. **87**, 203504 (2005)
20. Ichimura, S.: Current activities of ISO TC229/WG2 on purity evaluation and quality assurance standards for carbon nanotubes, Anal. Bioanal. Chem **396**, 963–971 (2010)

Section 3

21. Ichimura, S., Itoh, H., Fujimoto, T.: Current standardization activities for the measurement and characterization of nanomaterials and structures, J. Phys. Conf. Ser. **159**, 012001 (2009)
22. Itoh, H., Fujimo, T., Ichimura, S.: Tip characterizer for atomic force microscopy, Rev. Sci. Instrum. **77**, 103704 (2006)
23. Homma, Y., Takenaka, H., Toujou, F., Takano, A., Hayashi, S., Shimizu, R.: Evaluation of the sputtering rate variation in SIMS ultra-shallow depth profiling using multiple short-period delta layers, Surf. Interface Anal. **35**, 544–547 (2003)
24. http://www.npl.co.uk/nanoscience/surface-nanoanalysis/surface-and-nanoanalysis-research
25. Tougaard, S.: Surface nanostructure determination by x-ray photoemission spectroscopy peak shape analysis, J. Vac. Sci. Technol. **A14**, 1415–1423 (1996)

26. Hajati, S., Coultas, S., Blomfieldc, C., Tougaarda, S.: Nondestructive quantitative XPS imaging of depth distribution of atoms on the nanoscale, Surface Interface Anal. **40**, 688–691 (2008)
27. Baer, D., Amonette, J.E., Engelhard, M.H., Gaspar, D.J., Karakoti, A.S., Kuchibhatla, S., Nachimuthu, P., Nurmi, J.T., Qiang, Y., Sarathy, V., Seal, S., Sharma, A., Tratnyeke, P.G., Wang, C.-M.: Characterization challenges for nanomaterials, Surf. Interface Anal. **40**, 529–537 (2008)
28. Yacaman, M.J., Ascencio, J.A., Liu, H.B., Gardea-Torresdey, J.: Structure shape and stability of nanometric sized particles, J. Vac. Sci. Technol. B **19**, 1091 (2001)
29. Smith, D.J., Petfordlong, A.K., Wallenberg, L.R., Bovin, J.O.: Dynamic atomic-level rearrangements in small gold particles, Science **233**, 872 (1986)
30. Zhao, J.P., Chen, Z.Y., Cai, X.J., Rabalais, J.W.: Annealing effect on the surface plasmon resonance absorption of a Ti–SiO_2 nanoparticle composite, J. Vac. Sci. Technol. B **24**, 1104 (2006)
31. Wang, C.M., Baer, D.R., Amonette, J.E., Engelhard, M.E., Antony, J.J., Qiang, Y.: Electron beam-induced thickening of the protective oxide layer around Fe nanoparticles, Ultramicroscopy **108**, 43 (2007)
32. Jurac, S., Johnson, R.E., Donn, B.: Monte Carlo calculations of the sputtering of grains: enhanced sputtering of small grains, Astrophys. J. **503**, 247 (1998)
33. Gaspar, D.J., Laskin, A., Wang, W., Hunt, S.W., Finlayson-Pitts, B.J.: TOF-SIMS analysis of sea salt particles: imaging and depth profiling in the discovery of an unrecognized mechanism for pH buffering, Appl. Surf. Sci. **231–232**, 520 (2004)
34. Chen, H.H., Urquidez, O.A., Ichimura, S., Rodriguez, L.H., Brenner, M.P., Aziz, M.J.: Shocks in ion sputtering sharpen steep surface features, Science **310**, 294 (2005)
35. Gaspar, D.J., Engelhard, M.H., Henry, M.C., Baer, D.R.: Erosion rate variations during XPS sputter depth profiling of nanoporous films, Surf. Interface Anal. **37**, 417 (2005)
36. Jung, Y.J., Homma, Y., Vajtai, R., Kobayashi, Y., Ogino, T., Ajayan, P.M.: Straightening suspended single walled carbon nanotubes by ion irradiation, Nano Lett. **4**, 1109 (2004)
37. Baer, D.R., Engelhard, M.H., Gaspar, D.J., Matson, D.W., Pecher, K., Williams, J.R., Wang, C.M.: Challenges in applying surface analysis methods to nanoparticles and nanostructured materials, J. Surf. Anal. **12**, 101 (2005)
38. Zhang, H.Z., Gilbert, B., Huang, F., Banfield, J.F.: Water-driven structure transformation in nanoparticles at room temperature, Nature **424**, 1025 (2003)
39. Chernyshova, I.V., Hochella, M.F., Madden, A.S.: Size-dependent structural transformations of hematite nanoparticles. 1. Phase transition, Phys. Chem. Chem. Phys. **9**, 1736 (2007)
40. Chen, W., Pan, X.L., Willinger, M.G., Su, D.S., Bao, X.H.: Facile Autoreduction of Iron Oxide/Carbon Nanotube Encapsulates , J. Am. Chem. Soc. **128**, 3136 (2006)
41. Gliemann, H., Almeida, A.T., Petri, D.F.S., Schimmel, T.: Nanostructure formation in polymer thin films influenced by humidity, Surf. Interface Anal. **39**, 1 (2007)
42. Scher, E.C., Manna, L., Alivisatos, A.P.: Shape control and applications of nanocrystals, Philos. Trans. R. Soc. Lond. A **361**, 241 (2003)
43. Frankamp, B.L., Boal, A.K., Tuominen, M.T., Rotello, V.M.: Direct control of the magnetic interaction between iron oxide nanoparticles through dendrimer-mediated self-assembly, J. Am. Chem. Soc. **127**, 9731 (2005)
44. Karakoti, A.S., Kuchibhatla, S., Babu, K.S., Seal, S.: Direct synthesis of nanoceria in aqueous polyhydroxyl solutions, J. Phys. Chem. C **111**, 17232–17240 (2007)
45. Kuchibhatla, S., Karakoti, A.S., Seal, S.: Hierarchical assembly of inorganic nanostructure building blocks to octahedral superstructures – a true template-free self-assembly, Nanotechnology **18**, (2007)
46. Wertheim, G.K., Dicenzo, S.B.: Cluster growth and core-electron binding energies in supported metal clusters, Phys. Rev. B **37**, 844 (1988)
47. Dane, A., Demirok, U.K., Aydinli, A., Suzer, S.: X-ray photoelectron spectroscopic analysis of Si nanoclusters in SiO_2 matrix, J. Phys. Chem. B **110**, 1137 (2006)

48. Norman, T.J., Grant, C.D., Magana, D., Zhang, J.Z., Liu, J., Cao, D.L., Bridges, F., Van Buuren, A.: Near infrared optical absorption of gold nanoparticle aggregates, J. Phys. Chem. B **106**, 7005 (2002)
49. Reinhard, B.M., Siu, M., Agarwal, H., Alivisatos, A.P., Liphardt, J.: Calibration of dynamic molecular rulers based on plasmon coupling between gold nanoparticles, Nano Lett. **5**, 2246 (2005)
50. Bayer, M., Hawrylak, P., Hinzer, K., Fafard, S., Korkusinski, M., Wasilewski, Z.R., Stern, O., Forchel, A.: Coupling and entangling of quantum states in quantum dot molecules, Science **291**, 451 (2001)
51. Schwartz, D.A., Norberg, N.S., Nguyen, Q.P., Parker, J.M., Gamelin, D.R.: Magnetic quantum dots: Synthesis, spectroscopy, and magnetism of Co2+- and Ni2+-Doped ZnO nanocrystals, J. Am. Chem. Soc. **125**, 13205 (2003)
52. Liu, H., Brison, L.C.: A hybrid numerical-analytical method for modeling the viscoelastic properties of polymer nanocomposites, J. Appl. Mech. **73**, 758 (2006)
53. Glover, M., Meldrum, A.: Effect of "buffer layers" on the optical properties of silicon nanocrystal superlattices, Opt. Mater. **27**, 977 (2005)

Section 4

54. Lux research report 2004 on "sizing nanotechnology's value chain". http://www.luxresearchinc.com/pxn.php
55. Fujimoto, T., CIRJE-F-182 (Center for Intrnational Research on the Japanese Economy, Faculty of Economics, The Univ. of Tokyo): Architecture, capability, and competitiveness of firms and industries (2002)

Chapter 7
Implications of Measurement Standards for Characterizing and Minimizing Risk of Nanomaterials

David S. Ensor

7.1 Introduction

Nanotechnology as a concept is usually credited to Feynman [1] who presented the idea in a 1959 after-dinner speech entitled, "There's plenty of room at the bottom." Interest in nanotechnology at the national level grew to the point that the United States Government launched the National Nanotechnology Initiative (NNI) in 1999 [2]. From a programmatic standpoint, materials related disciplines were combined using the unifying principle that some feature of the material should fall within the nanoscale size range. Nanoscale is defined as the size from approximately 1–100 nm [3]. Also some well-known materials associated with nanotechnology, such as fullerene and single wall carbon nanotubes were discovered in only the last 25 years [4, 5]. Much of the supporting science is well established in fields such as electronics, polymers, powders, colloids, and aerosols. However, the nanotechnology field is currently expanding rapidly with the discovery of new techniques, insights, applications and materials. It is clear that unifying principles and appropriate standards need to be developed to allow a systematic approach to managing the applications and risks of nanotechnology. These challenges have been faced by ISO Technical Committee 229 "Nanotechnologies" in its program to develop documents consistent with the goals of international standardization. The purposes of international standardization are to facilitate international trade; improvement of quality, safety, security, environmental and consumer protection, as well as the rational use of natural resources; and global dissemination of technologies and good practices [6].

This chapter explores standards development and the implications on how nanomaterial measurement standards might be used to achieve international goals. This activity will require reconciliation of the properties of the newly developed materials and the emerging health concerns about potential large scale industrial applications.

D.S. Ensor (✉)
RTI International, Research Triangle Park, Durham, NC, USA
e-mail: dse@rti.org

V. Murashov and J. Howard (eds.), *Nanotechnology Standards*,
Nanostructure Science and Technology, DOI 10.1007/978-1-4419-7853-0_7,

In many cases the nanoscale material may result from a process change, for example, to reduce the particle size of a powder to achieve properties desirable for new applications. Or in some cases, the materials may have been recently synthesized and are available only in test quantities. The potential environmental, health and safety (EHS) concerns associated with of these new materials have been discussed by Oberdörster et al. [7], Borm et al. [8], and Maynard and Kuempel [9]. The approach of using risk assessment to understand the implication of the manufacture and use of nanoscale products will be an important part of international nanomaterial risk management.

7.2 Risk Paradigm

Since a major thrust of this chapter is examining measurement standards and implications of these how to minimize risk of nanomaterials, it is useful to examine current risk assessment concepts [10]. The risk paradigm was initially formulated by the US National Academy of Sciences [11] and has been used to guide environmental programs for many years. The risk paradigm and associated tools such as life cycle analysis are believed to be directly applicable to the assessment of nanomaterials [12].

Shown in Fig. 7.1 is a high level diagram of the risk assessment and risk management paradigm. The risk paradigm is a systematic process to identify, quantify and to set priorities to manage risk. The four elements of risk assessment can be quite complex when population statistics and the multitude of methodologies

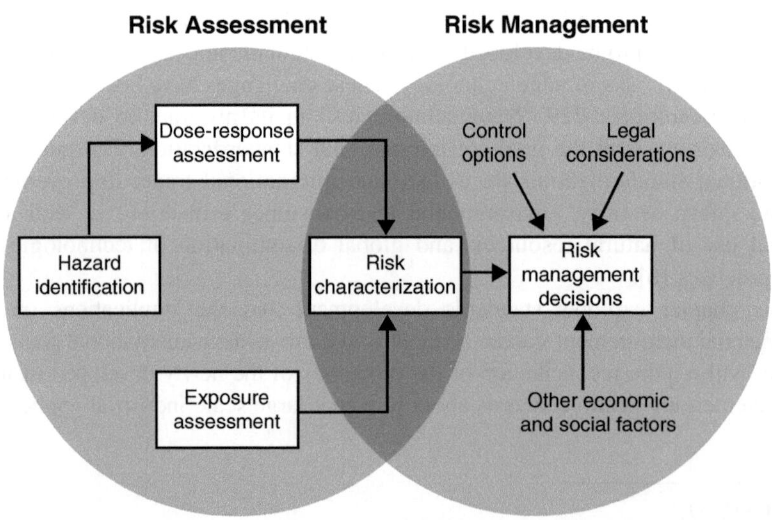

Fig. 7.1 Diagram of the risk paradigm showing the general elements for the management of toxic materials in the environment [10]

available to develop this information are considered. Hazard identification involves identifying the adverse effects that may occur from exposure to environmental stressors or agents. Also the characterization of the nature and strength of these stressors need to be determined. After the hazard has been identified, the exposure of populations to the material maybe assessed. Exposure assessment involves describing the population or ecosystems exposed to stressors and the magnitude, duration, and spatial extent of exposure. As hazards are identified, efforts are made to determine the dose response characteristics or toxicity of the material which may be determined in vitro, in vivo or by epidemiological studies. Based on both exposure and dose-response properties, the risk of the material is characterized. Underlining each step of risk assessment is the quantification of the amount and the characteristics of the materials. When the risk is quantified then management processes can be applied to minimize risk. Later in this chapter, three different ISO/TC 229 project activities are summarized as examples to illustrate how international measurement standards may support various steps in the risk paradigm.

7.3 Nanotechnology Standards Development

Hatto [13] described the mission of ISO/TC 229 in developing standards in light of the short history of the field. The fact that the standards are technology based rather than narrowly science or application based means that: (1) close collaboration and liaisons will need to be established with other standards committees and organizations, (2) standards developed for nanotechnology need to build on and augment existing standards, and (3) standards developed by ISO/TC 229 can provide normative authority in ISO standards written by other committees. Often when an ISO technical committee is started, the organizers have a large inventory of long standing national standards that are ready for harmonization at the international level. With few exceptions, this has not been the case for nanotechnology.

An important consideration is that nanotechnology standards will build heavily on standards previously developed by other technical committees and organizations. This is particularly true for metrology and EHS standards. These existing documents may have information relevant to the nanoscale but it may be not identified as such. Often existing standards may have some elements that are important to provide the context of a new standard with a nanotechnology focus.

Relevant international standards were developed before the organization of nanotechnology as a distinct field and might be imperfectly applied to support nanotechnology standards. There is always a question on how to best apply this information in drafting a new standard. In some cases for example, the existing standard may not apply to smallest particle diameters in the nanoscale size range. Another possibility is that an existing standard may never have been actually used for the nanoscale materials of interest. Many of the existing standards in the powder technology area were developed for powders with a small fraction of nanoparticles. Therefore a question that must be asked – Can performance established in the micro-scale range be extrapolated to particles in the nanoscale range? Often for this reason, some of the projects undertaken by ISO/TC 229 may have a research orientation rather than representing settled practice.

The literature base of direct nanoscale applications may be small for a particular area of interest. Only a few researchers may be actively working in the area and often the research is only now underway. This situation has caused a tension in the writing of documents because of the limited number of experts and the vague feeling that the approach selected may be incomplete. However, this activity in nanotechnology standards will serve to identify areas where existing practice can be accelerated. The standards development process in ISO is robust with respect to developing standards in new areas. ISO standards are systematically reviewed initially within 3 and 5 years thereafter and can be updated to include contemporary experience.

A common feature in the application of standards to nanotechnology is the possibility to develop nanotechnology content in ISO/TC 229 standards that can be widely applied by other standards committees. Potentially many documents in topics such as terminology, measurement, material specifications, and health, safety and environment developed in ISO/TC 229 may have wide application in support of other ISO standards. Nanotechnology is expected become an important factor in a large number of fields because of its enabling nature.

An example of the relationship of existing standards and new standards developed in response to a nanotechnology is illustrated by current standards activity in ISO/TC 209 "Cleanrooms and associated controlled environments" [14]. When ISO/TC 209 was organized in 1993, contamination control practice was generally restricted to particles 100 nm and larger. One of the factors in that decision was a rule-of-thumb used in the electronics industry that only particles larger than 1/10 of a circuit line width could cause damage during manufacture of a semiconductor chip. At that time, circuit line widths in electronics were about 1 μm naturally leading to lower particle size limit of 100 nm in the standards. As a result the first cleanroom standard ISO 14644-1 [15], excluded sub 100 nm particles with the exception that ultrafine particle concentration as measured with a condensation particle counter could be reported in addition to certification data taken with an optical particle counter. Ultrafine particle is an older term for particles less than 100 nm without identity with respect to source. However, most of the elements of contamination control such as design for separation of the work areas from the environment and operations management are independent of the contaminating particle size. Working Group 10 in ISO/TC 209 is examining the extension of the committee's standards into particle size ranges important for nanotechnology. Currently the semiconductor industry now makes circuits of smaller than 65 nm [16] feature size, clearly the particle size range of interest in contamination control will need to be reduced in new standards. Much of the material developed in ISO/TC 229 for terminology, and EHS will have direct use in the ISO/TC 209 nanotechnology standards.

7.4 Relating Measurement Standards to the Risk Paradigm

Three examples of the standards under development in ISO/TC 229 were selected to illustrate aspects (exposure assessment, dose-response assessment and hazard identification) of the risk paradigm in Fig. 7.1. The first example is a standards project to determine the quantity of nano-objects that can be aerosolized from powders. Nano-object is a more general term with respect to shape and includes nanoparticles, nanofibers, and nanoplates [3]. The term nanoparticle is used for the case when all three dimensions are of the nanoscale. (Some of the older documents tend to use nano-object and nano-particle interchangeably. Particle is an even more general term because it is usually not limited to a specific size range in most definitions.) This planned standard is intended to characterize the exposure potential of powders when handled. The second example is a pair of standards describing the generation and characterization of silver nanoparticle aerosol for inhalation studies to support the dose-response assessments. The third and final example is the measurement of endotoxin associated with manufactured carbon nanomaterials. This standard is in effect a component of hazard identification because although the health effects of endotoxin dose have been well established in other fields, the presence of endotoxins in manufactured nanomaterials has not been previously identified as a concern.

7.4.1 Nano-object Content of Powders

This standards development project addresses one aspect of hazard identification in the risk paradigm shown in Fig. 7.1. This project is ISO/CD 12025 *Nanomaterials – General framework for determining nanoparticle content in nanomaterials by generation of aerosols* [17] and is currently at the committee draft stage [18]. There has been discussion on the appropriateness of the title because the test method is used to determine the quantity of nano-objects aerosolized or liberated from the powdered material and not the total concentration of nano-objects in a specified size found in the powdered material.

One of the issues motivating development of this standard is ability of powders to liberate potentially hazardous particles during handling operations in manufacturing processes. The quantity of aerosolized nano-objects as determined by this standard, in effect related to the likelihood of exposure would need to be combined with a measurement of toxicity to estimate dose in order to assess the potential risk.

Traditionally the laboratory "dustiness" tests of powdered materials have been used by the industrial hygiene community to determine the extent of engineering controls that are required to safely handle or process the materials. Dustiness measurements consist of subjecting a powder in a bench-scale test designed to simulate handling in an industrial operation. The liberated dust is sampled to determine the respirable fraction by mass as a ratio to the initial mass of powder. As described by Hinds [19] respirable mass is determined by sampling the aerosol through a cyclone designed with a 50% cut point at a particle diameter of 3.5 μm, mimicking the human upper respiratory system. The mass expected to reach the alveolar region of the lung is determined by sampling the outlet of the cyclone with a filter. Other respiratory tract based size fractions often used include the inhalable particle size with a cut-off of 15 μm and the thoracic size with a 2.5 μm particle sizecut-off. Over the last 50 years, a large number of these laboratory tests have been developed [20]. Although the potential tests are empirically based, effort should be taken to prevent confounding factors such as the effects of relative humidity [21].

All the aerosolization methods apply mechanical energy in some manner to a dry powder while air flows through the system to transport the liberated aerosolized particles to instruments for measurement. The approach taken in writing ISO/CD 12025 is to indicate the key properties of the powder influencing the outcome, safety, instrumentation and examples of methods to disseminate the powder. However, a single method was not specified because of the wide range of powders of potential interest. This approach is illustrated in Fig. 7.2. The energy may be applied to the powder by a wide range of methods including: tumbling the powder in a drum, dropping the powder in a tube from a height, vibration of the powder in a fluid bed, or subjecting the powder to a pressure pulse of air. The tumbling drum and the drop tube methods are described in a CEN standard [22]. The vibration-fluid bed method consists of placing the powder either with or without bronze beads about 70 μm in diameter in a glass vial attached to a laboratory vortex shaker. Air is aspirated through the vial and the aerosolized particles are transported to aerosol measurement instrumentation [23]. In the pressure pulse or dynamic method a small quantity of dust is resuspended and transported by a blast of air to a small chamber for measurement with instruments a much lower flow rate. The dynamic method is particularly interesting because of the small quantity of dust required (5 mg) and the ability to contain the dust in a chamber [24].

Quantification of the aerosolized particles will likely require a combination of simultaneous test methods. The traditional respirable dust fraction should be determined to maintain continuity with the dustiness data base. However, in addition measurements should be made of the nano-particle concentration as a function of particle size as well. The available test methods to measure nanoparticles in the workplace have been summarized in ISO/TR 27628 [25] developed by ISO/TC 146. These methods include:

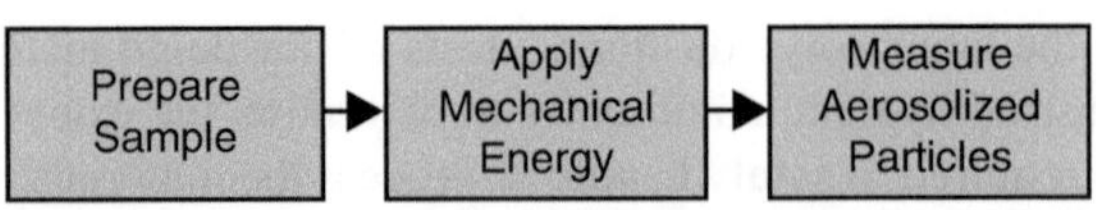

Fig. 7.2 Simplified schematic of the aerosolization process

- Scanning mobility particle analysis
  The particles are separated by size by electrical charge in a differential mobility analyzer and detected with optical scattering by condensing fluid on the particles.
- Electrical low pressure impactor
  The particles are separated by size from a series of jets with ever increasing velocities directed at flat surfaces or stages and are impacted. The rear part of the instrument is operated at low pressure reducing drag on the particles allowing deposition of particles smaller than 100 nm. Detection is accomplished by measuring the electrical charge on the particles collected on impaction stages corresponding to specific particle diameters.
- Sampling onto membrane filters with subsequent electron microscopy
  This is a labor intensive two step process and for quantification requires measurement of a statistically valid number of particles. The images of the particles will provide information on the shape of the particles and state of aggregation. The particle sizes and experimental purpose will dictate the selection of electron microscope, e.g. scanning electron microscope or transmission electron microscope.

However, at the present time, consensus has not been reached on the appropriate measurement methodology for liberated nano-objects and the best approach to interpret the results. Differential mobility analysis described in ISO 15900 [26] developed by ISO/TC 24 for particle size separation combined with condensation particle counting for detection is a widely used to determine the nanoparticle size distribution. It is also desirable to perform parallel sampling with respirable mass sampling methods in addition to nanoparticle measurements to obtain comparability with data obtained with conventional dustiness test methods.

However, very little information is available on the aerosolization of nanoparticles from powders in a dustiness context. Only a few laboratories world-wide have experience with nano-objects and that data is only now being published in the open literature. Some of the limited data has been summarized by Schneider and Jensen [27] demonstrating that the particle size number distributions exhibit distinct modes. It was also reported that high shear methods such as the ELPI may cause deagglomeration during sampling.

Because of the generic nature of the standard, the requirements sections focus on selection of the appropriate test approaches, documenting the test conditions and data reporting. The results oriented approach is supported by the ISO Directives [28]. The generic approach might allow application to a wider range of powders than if a specific method was selected. There is however the serious question when comparing data taken with different dustiness methods.

It is expected that publication of this standard will stimulate work in this area. In particular if regulatory activities require dustiness or nanoparticle aerosolization information. There may also be a possibility that additional new standards might be written around specific techniques of aerosolizing powders.

7.4.2 Metal Aerosol Inhalation Standards

This standard provides methodology for inhalation testing as part of research to obtain data to support dose-response assessments in the risk paradigm in Fig. 7.1. Nanoscale silver has become widely used as an antimicrobial in a large number of products, for example see Woodrow Wilson International Center for Scholars [29]. Chen and Schluesener [30] reviewed the growing and widespread used of nanoscale silver in a number of medical applications. Quadros and Marr [31] reviewed the environmental and human health risks and concluded that inhalation exposure is of the greatest concern because of numerous opportunities for aerosolization during nanoscale silver's product lifecycle. Also, inhalation is a way to deliver nanomaterials to test animals to understand the transport and accumulation of the nanoparticles within the organs of test animals.

A widely used method of generating silver aerosol is by thermal generation [32]. Thermal generation involves heating silver or gold metal to volatize the metal and nanoparticles are formed when the gas containing metal atoms are cooled by mixing with air. Traditionally, a tube furnace is used to heat a ceramic boat containing the silver, gas is piped through the furnace to transport the volatilized metal and then the gas is mixed with cold air to form the aerosol. Jung et al [33] described a simpler method where the silver metal is placed on a small ceramic heating element and air is allowed to flow over the element cooling and transporting the aerosol. The resulting aerosol containing silver nanoparticles is characterized and used to expose laboratory animals [34]. The steps are shown in Fig. 7.3. These two standards, ISO/DIS 10801 [35] and ISO/DIS 10808 [36] are currently at the FDIS stage [18].

The standards were based on aerosol generation method reported by Ji et al. [34] and characterization parts of the standards were derived from established principles of aerosol measurement such as in ISO 15900 and inhalation sections were derived from OECD Guidelines [37]. However these OECD exposure guides were originally written for chemical vapor exposure but generally cover the requirements for the chambers independent of test substance such as air exchange rates, oxygen levels, and temperature and relative humidity tolerances.

When the work was started on writing the standard, only one laboratory had experience on the animal exposure method. Fortunately, the technical approach was supported by several peer reviewed publications. It is believed that with publication of the standard that it will be used by a number of laboratories.

The key elements relevant for nanoparticle exposure were referenced in the final standard. In both standards, example experimental set-ups and data are given in the annex to guide the researcher. ISO 10801 is organized around the steps of preparing

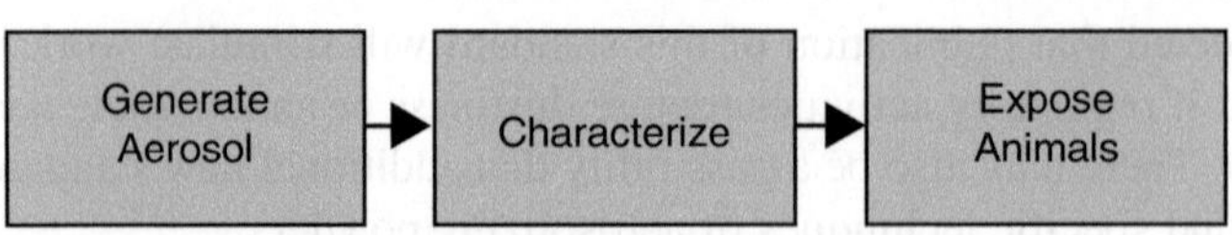

Fig. 7.3 Schematic of silver aerosol generation and characterization

the generation system, the characterizing the aerosol generator, requirements for particle generation, assessment of results and the test report. Ideally, a range of aerosol generation equipment is permitted if the basic requirements of size distribution properties and concentration stability are followed. ISO 10808 focuses on the monitoring of the animal chambers. The standard is intended to be a companion to ISO 10801 but because of its generic requirements based structure, the document could be applied to other aerosol generators as well. The standard follows the steps of experimental program execution following preparation of the system, specification of the monitoring methods, assessment of results and the test report.

Both ISO 10801 and ISO 10808 are expected to be issued as standards in 2011. Publication of these standards will establish precedence for inhalation studies of small animals. These documents are expected to form a cornerstone for this area of research.

7.4.3 Quantification of Endotoxin in Nanomaterials

The endotoxin in nanomaterials was identified as a potential consideration in the exposure assessment step in Fig. 7.1. The importance of endotoxins in the toxicology of nanomaterials is only now being recognized. In addition, manufactured nanomaterials or nano-objects often have large specific areas, and could potentially adsorb endotoxin from the environment. Concentrated endotoxin might be carried by nano-objects to biological sites in a manner similar to mechanisms proposed for nanotechnology enabled therapeutics.

There is very little literature on endotoxin levels in manufactured nanomaterials. The toxicity of nano-objects containing endotoxin is important for several reasons: (1) during manufacture workers might be exposed to the material, (2) during the life cycle of the nanomaterial in the environment endotoxins might be accumulated and (3) many of the nano-objects might be used as precursors for therapeutics, and (4) often commercial nano-objects are used in toxicology studies without consideration of all the potential confounding factors. The presence of endotoxin is well known to cause health problems in the indoor environment. The cell walls of gram-negative bacteria contain endotoxin and these bacteria are wide spread and endotoxin is widely distributed in the environment.

Injectables and medical devices must be screened for endotoxin as required by National Pharmacopeias. An assay is well established based on Limulus amebocyte lysate (LAL) and is available as a test kit [38]. A potential confounding factor caused by environmental endotoxins on interpreting toxicity data obtained with nanomaterials was identified by Inaba [39]. Puzzling results were discovered when commercial carbon nanotubes were subjected to in vitro testing. It was believed that endotoxins were confounding in vitro toxicology studies of nanoparticles. In 2006, an ISO/TC 229 standards writing effort was started to write an international standard [40] covering the measurement of endotoxin on nanomaterials. Carbon nanomaterials for example are made by processes that involve elevated processing temperatures in chemical vapor deposition or electric arc reactors. Just after synthesis, the freshly made materials are quite likely endotoxin free. The endotoxin contamination could only have been introduced in the purification and storage process. This possibility was also recognized in the United States with the publication of an endotoxin test method for nanomaterials in in vitro dosing solutions [41]. However, if a nanomaterial is used for toxicity tests in the dry state, the presence of endotoxin is typically not considered.

The ISO standard written summarizes the current way of measuring endotoxins using the LAL assay. It provides minimum requirements for quantifying endotoxin on nanomaterials. However, considering the wide range of possible materials only guidance is given with respect to analysis of nanomaterials. Laboratories will need to develop a method for their particular materials. Validation will need to be made of the particular material evaluated to ensure the results are quantitative. The standard was written using existing pharmacopeias test requirements with guidance on analyzing nano-objects which are often hydrophobic and very difficult to suspend in water based systems. An addition concern was that the nanomaterials themselves might interfere with the assay. One of the ways of measuring endotoxin is to measure the change in light extinction since the nano-objects are often aggregated particles the material might cause interference with the sample. Sample preparation will still need to be developed and validated for every class of material.

During the time that the standard was being written, RTI International conducted a short internally funded study reported by Esch et al. [38] to develop a method for measuring endotoxin associated with dry carbon nano-objects such as: single wall nanotubes, multiwall nanotubes, fullerene C_{60} and with carbon black as a control. Preliminary data were obtained on the levels of endotoxin in a small number of commercial materials as shown in Fig. 7.4. Nanomaterials pose a problem in current assays in that the materials are provided as hydrophobic dry powders which are very difficult to directly suspend in the LAL regent. It is desirable to screen the dry materials as received from the manufactures to identify if the material contains endotoxins in particular if the materials is aerosolized for inhalation studies. As described in Esch et al. [38] after trying several different surfactants, Vitamin E d-α-tocopheryl polyethylene glycol – 1000 succinate (VETPGS) surfactant was found to provide excellent sample preparation. Typically a 1% solution of VETPGS solution in endotoxin-free water was used in the preparation. The published paper [38] was referenced for guidance in the bibliography of the standard [40] for the guidance of users. VETPGS wets the carbon nanomaterial making suspension of the materials feasible for instrumental analysis. In results reported by the Esch et al. [38], the concentration of endotoxin was not correlated to the specific surface of the materials but appeared to be randomly introduced during purification of the materials. This endotoxin contamination was not introduced in the laboratory analysis because the dry material was removed from the shipping containers in an inert atmosphere glove box. In addition, the endotoxin analysis was performed with precautions to avoid contamination. The Esch et al. [38] paper found that the samples analyzed would have caused an adverse reaction based on the regulatory limits of endotoxin.

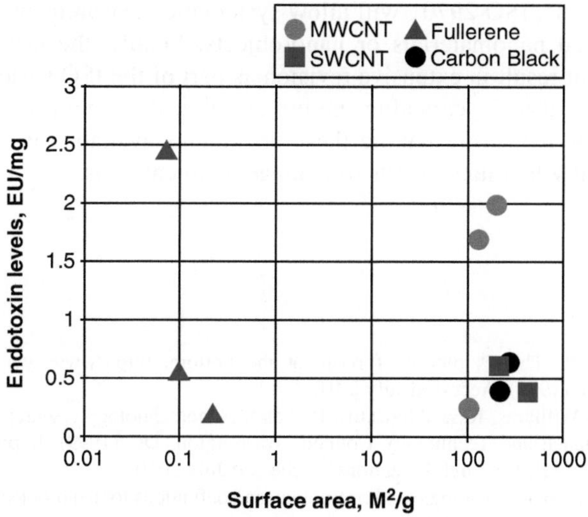

Fig. 7.4 Endotoxin contamination as a function of surface area for a number of manufactured carbon nanomaterials. Reproduced from Esch et al. [38]

ISO/FDIS 29701 is current in final ballot and expected to be a completed standard later in 2010. It will be the first international standard developed by ISO/TC 229. In summary, this standard will provide the basis of systematic investigation of endotoxins in dry manufactured nanomaterials. It is believed as the toxicity testing of nanomaterials becomes more systematic and developed that this test method will be used widely.

7.5 Summary

This chapter explored implications of three measurements standards in terms of the risk paradigm. The standards under development by ISO/TC 229 are important in developing data by approaches that have been developed by the consensus process. Since the nanotechnology field is new, the standards need to draw heavily from other fields. Many of these fields can be considered to have formed the basis of nanotechnology. However, some of the supporting standards may need some adaptation for application in the nanotechnology context. In each of all three examples only a small number of laboratories had experience with the analytical methods. It is believed that all three will form an important part of the standards literature. ISO 12025 will quite likely provide a generic basis for determining dustiness of powders containing nano-objects. It is expected that this standard will be used widely to support both industrial hygiene and environmental applications to determine potential exposure associated with various powders. ISO 10801 and 10808 will quite likely establish the basis for aerosol inhalation testing with nano-objects. ISO 29701 will allow systematic measurements of endotoxin in manufactured nanomaterials or nano-objects. Finally, the newness of these standards might result in extensive revision as part of the ISO systematic review, the first is scheduled 3 years after the initial publication and every 5 years after the first review. However, making these documents available on an accelerated basis will greatly facilitate efforts to manage nanomaterials.

References

1. Feynman, R.P.: There's plenty of room at the bottom. http://www.zyvex.com/nanotech/feynman.html (1959). Accessed July 2010
2. Roco, M.C., Williams, R.S., Alivisatos, P. (eds.): Nanotechnology research directions. U.S. National Science and Technology Council, Washington, DC (1999). http://www.wtec.org/loyola/nano/IWGN.Research.Directions/. Accessed July 2010
3. ISO TS 27687. Nanotechnologies – Terminology and definitions for nano-objects – nanoparticle, nanofibre and nanoplate (2008)
4. Kroto, H.W., Heath, J.R., O'Brian, S.C., Curl, R.F., Smalley, R.E.: C_{60}: Buckminsterfullerene. Nature **318**, 162–163 (1985)
5. Iijima, S.: Helical microtubules of graphitic carbon. Nature **354**, 56–58 (1991)
6. ISO strategic plan. http://www.iso.org/iso/isostrategies_2004-en.pdf. Accessed July 2010
7. Oberdörster, G., Maynard, A., Donaldson, K., Castranova, V., Fitzpatrick, J., Ausman, K., Carter, J., Karn, B., Kreyling, W., Lai, D., Olin, S., Monteiro-Riviere, N., Warheit, D., Yang, H.: Principles for characterizing the potential human health effects from exposure to nanomaterials: elements of a screening strategy. Part. Fiber Toxicol. **2**, 8 (2005). http://www.particleandfibretoxicology.com/content/2/1/8. Accessed July 2010
8. Borm, P.J.A., Robbins, D., Haubold, S., Kuhibusch, T., Fissan, H., Donaldson, K., Schins, R., Stone, V., Kreyling, W., Lademann, J., Kertmann, J., Warheit, D., Oberdörster, E.: The potential risks of nanomaterials: a review carried out for ECETC. Part. Fibre Toxicol. **3**, 11 (2006). http://www.particleandfibretoxicology.com/content/3/1/11. Accessed July 2010
9. Maynard, A.D., Kuempel, E.D.: Airborne nanostructured particle and occupational health. J. Nanopart. Res. **7**, 587–614 (2005)
10. U.S. Environmental Protection Agency. The NRC risk assessment paradigm. http://www.epa.gov/ttn/atw/toxsource/paradigm.html. Accessed July 2010
11. U.S. National Academy of Sciences. Assessment in the Federal Government: Managing the Process. National Academy Press, Washington, DC (2008). http://books.nap.edu. Accessed July 2010
12. U.S. Environmental Protection Agency. Nanotechnology white paper, EPA 100/B-07/001, February 2007. http://www.epa.gov/osa. Accessed July 2010
13. Hatto, P.: Nanotechnologies – ISO/TC 229, ISO, IEC, NIST and OECD International workshop on documentary standards for measurement and characterization for nanotechnologies, NIST, Gathersburg, MD, 26–28 February 2008
14. Dixon, A.M., Ensor, D.S., Michael, D.: Applying the principles of contamination control standardization to nanotechnology facilities, IESC 2010, Tokyo, Japan, 6–9 October 2010
15. ISO 14644-1. Cleanrooms and associated controlled environments – Part 1: classification of air cleanliness (1999)
16. Moore's Law 40th Anniversary. http://www.intel.com/pressroom/kits/events/moores_law_40th/ (2005). Accessed July 2010
17. ISO/CD 12025. Nanomaterials – General framework for determining nanoparticle content in nanomaterials by generation of aerosols

18. ISO ISO/TC 229. http://www.iso.org/iso/iso_catalogue/catalogue_tc/catalogue_tc_browse.htm?commid=381983&development=on (2010). Accessed July 2010
19. Hinds, W.C.: Aerosol Technology. Wiley, New York, NY (1982)
20. Hamelmann, F., Schmidt, E.: Methods of estimating the dustiness of industrial powders – A review. KONA **21**, 7–18 (2003)
21. Pinke, M.A.E., Leith, D., Boundy, M.G., Loffler, F.: Dust generation from handling powders in industry. Am. Ind. Hyg. Assoc. J. **56**, 251–257 (1995)
22. EN 15051. Workplace atmospheres – measurement of the dustiness of bulk materials – requirements and test methods (2006)
23. Maynard, A.D., Baron, P.A., Foley, M., Shvedova, A.A., Kisin, E.R., Castranova, V.: Exposure to carbon nanotube material: Aerosol release during the handling of unrefined single-walled carbon nanotube material. J. Toxicol. Environ. Health A **67**, 87–107 (2004)
24. Boundy, M., Leith, D., Polton, T.: Method to evaluate the dustiness of pharmaceutical powders. Ann. Occup. Hyg. **50**(5), 453–458 (2006)
25. ISO/TR 27628. Workplace atmospheres – Ultrafine, nanoarticle and nano-structured aerosols – inhalation exposure characterization and assessment (2007)
26. ISO 15900. Determination of particle size distribution – Differential electrical mobility analysis for aerosol particles
27. Schneider, T., Jensen, K.A.: Relevance of aerosol dynamics and dustiness for personal exposure to manufactured nanoparticles. J. Nanopart. Res. **11**, 1637–1650 (2009)
28. ISO/IEC. Directives, Part 2: Rules for the structure and drafting of International Standards. http://isotc.iso.org/livelink/livelink?func=ll&objId=4230456&objAction=browse&sort=subtype. Accessed July 2010
29. Woodrow Wilson International Center for Scholars. A Nanotechnology Consumer Products Inventory. Washington, DC (2010). http://www.nanotechproject.org/inventories/consumer/. Accessed July 2010
30. Chen, X., Schluesener, H.J.: Nanosilver: A nanoproduct in medical application. Toxicol. Lett. **176**, 1–12 (2008)
31. Quadros, M.E., Marr, L.C.: Environmental and human health risks of aerosolized silver nanoparticles. J. Air Waste Manag. Assoc. **60**, 770–781 (2010)
32. Ku, B.K., Maynard, A.D.: Comparing aerosol surface-area measurement of monodisperse ultrafine silver agglomerates using mobility analysis, transmission electron microscopy and diffusion charging. J. Aerosol Sci. **36**, 110–1124 (2005)
33. Jung, J.H., Oh, H.C., Noh, H.S., Ji, J.H., Kim, S.S.: Metal nanoparticle generation using a small ceramic heater with a local heating area. J. Aerosol Sci. **37**, 1662–1670 (2006)
34. Ji, J.H., Jung, J.H., Kim, S.S., Yoon, J.U., Park, J.D., Choi, B.S., Chung, Y.H., Kwon, I.H., Jeong, J., Han, B.S., Shin, J.H., Sung, J.H., Song, K.S., Yu, I.J.: Twenty-eight-day inhalation toxicity study of silver nanoparticles in Sprague Dawley Rats. Inhal. Toxicol. **19**(10), 857–871 (2007)
35. ISO/DIS 10801. Nanotechnologies – Generation of metal nanoparticles by evaporation/condensation method for inhalation toxicity testing
36. ISO/DIS 10808. Nanotechnologies – Characterization of nanoparticles in inhalation exposure chambers for inhalation toxicity testing
37. OECD: Guidline for Testing of Chemicals 413 Subchronic Inhalation Toxicity: 90-Day Study. OECD, Paris (1995)
38. Esch, R.K., Han, L., Ensor, D.S., Foarde, K.K.: Endotoxin contamination of engineered nanomaterials. Nanotoxicology **4**, 73–83 (2010)
39. Inaba, K.: Standardization of endotoxin test. Presented at the 2nd ISO/TC229 Plenary Meeting as document TC229/N149, Tokyo, Japan, June 2006
40. ISO/FDIS 29701. Nanotechnologies – Endotoxin test on nanomaterial samples for in vitro systems – Limulus amebocyte lysate (LAL) test
41. Nanotechnology Characterization Laboratory, NCL Method STE-1. http://ncl.cancer.gov/NCL_Method_STE-1.pdf. Accessed July 2010

Chapter 8
Nanomaterial Toxicity: Emerging Standards and Efforts to Support Standards Development

Laurie E. Locascio, Vytas Reipa, Justin M. Zook, and Richard C. Pleus

8.1 Introduction

For the first time in the history of industrialization, nanotechnology offers the unique opportunity to consider material safety concerns prior to widespread adoption and use by industry. Many scientists around the world have been motivated by this and are working on developing and applying nanotechnology as safely as possible, attempting to avoid the pitfalls of our earlier introductions of new chemicals and chemical processes into commerce. One key aspect of defining the safety of any chemical product, whether nano-sized or conventional, is toxicity testing and the determination of hazard potential during manufacturing and/or use.

Toxicity testing, simply stated, is scientifically based testing to determine the potential toxic effects of a product or any and all components of a product. The testing of these can be conducted using a number of methods, including *in vitro*, *in vivo*, and epidemiological, and the results of these tests feed into risk assessment of the material. Assessing risks of products to workers, consumers, or the environment is a well-established process initially defined by the US National Academies of Science in 1983 [1]. Today, risk assessment has evolved and is practiced world-wide. In general, nanomaterial toxicity testing should follow similar protocols to conventional chemical testing; however, what we have learned so far is that there are likely some important differences between nanomaterials and typical chemicals that will factor into testing protocols [2]. While chemical structure is the most important consideration when testing conventional chemicals, nanomaterials are much more complex structures where multiple physico-chemical characteristics are likely to play a key role in defining a nanomaterial's potential toxicity and hazard. For example, researchers have reported relationships between nanomaterial toxicity and parameters such as size [3, 4], shape [5], aggregation state [6, 7],

L.E. Locascio (✉)
Biochemical Science Division, National Institute of Standards and Technology, Gaithersburg, MD 20899-8310, USA
e-mail: locascio@nist.gov

V. Murashov and J. Howard (eds.), *Nanotechnology Standards*, Nanostructure Science and Technology, DOI 10.1007/978-1-4419-7853-0_8,

and surface chemistry [8]. Additionally, the toxicity of nanoparticles may be greater than that of the bulk material due to their larger reactive surface area-to-mass ratio, faster dissolution, and ability to enter into cells resulting in the "Trojan horse mechanism," where the particles have a particularly toxic effect inside the cells [9, 10]. This is more closely linked to our experience with the stereo-chemistry of receptor-ligand interactions in pharmacology where physico-chemical characteristics including size, shape, exposed chemical moieties, and spatial orientation (handedness) greatly influence potentially toxic effects. Much work is being conducted to ensure that physico-chemical characteristics of nanomaterials are assessed prior to toxicological testing in an attempt to establish linkages between material properties and toxicology [11, 12].

Also unique to nanomaterial toxicity testing are complications related to the measure of dose, defined as material concentration versus administration time. Currently, there is no internationally accepted measure of concentration or dose for toxicity testing of nanomaterials, and dosage is commonly expressed in terms of particle number, mass, and/or surface area. Because of the small size of nanomaterials, the surface area expressed dosage can be quite large while the mass expressed dosage can be very small. Therefore, conclusions as to whether it is toxic at a high dosage (expressed as surface area) or low dosage (as expressed by mass) can be quite confusing. As Paracelsus, the father of toxicology, stated, "…all substances are poisons, there is none that is not…it is the dose that differentiates a poison from a remedy." Therefore, coming to consensus on the measure of dosage for nanomaterial toxicity testing is critical.

An additional consideration unique to nanomaterial toxicology is the potential ability of nanomaterials to be absorbed into organisms, organs, tissues and individual cells via previously unexpected routes. For instance, manganese oxide nanoparticles have been demonstrated in monkeys to be transported via the olfactory nerve to the brain [13]. Other work has demonstrated that lung macrophages remove nanoparticles less efficiently than larger particles, and that nanoparticles can be transported to many tissues and organs through the circulatory, lymphatic, and nervous systems [14].

Due to the rapid product development of nanomaterials and applications currently underway, it is imperative that we understand the toxicity of nanomaterials as rapidly as possible in order to support product development timelines. How do we attempt to bring nanotechnology safely into commerce in the fastest way possible on a global level? One way is to establish a universal approach to understanding the environmental, health and safety effects of nanomaterials through science-based and consensus-based written standard methods (also known as documentary standards) for their toxicity testing and risk assessment. What started in the early to mid-2000s was the beginning of the development of documentary standards for nanotechnology and, related to this chapter, for nanotoxicology.

Early in the 2000s, and preceding the international nanotechnology standards efforts, was the establishment of national and regional standards efforts related to nanotechnology. China was the first to establish a national standards effort in nanotechnology in December 2003 with the United Working Group for Nanomaterials

Standardization. Efforts in the UK, the USA, and Japan followed in 2004 with the establishment of the British Standards Institute (BSI) Committee for Nanotechnologies – NTI/1 in the UK, American National Standards Institute (ANSI) Nanotechnology Standards Panel in the USA, and the Committee for Nanotechnology Standardization Research and Study in Japan. The first regional standards effort in this field was launched in 2005 with the establishment of CEN TC/352 European Committee for Standardization Nanotechnologies. Also in 2005 came the first international standards efforts within the International Organization for Standardization (ISO) and ASTM International. As the nanotechnology standards field has evolved, many of the national and regional activities are now focused on providing coordinated input into the international efforts in ISO, ASTM and the Organization for Economic Cooperation and Development (OECD) rather than developing independent national standards. The activities in many nationally organized programs specifically related to nanotoxicology are also fed into these global standards development organizations (SDOs), thereby effectively coordinating standards related to nanotoxicology around the world. In this way, this special field of nanotoxicology is becoming highly organized and coordinated with a very dynamic international dialogue, and this is having a profound and revolutionary impact on the field of toxicity testing in general. A list of national standards efforts that have some specific focus on nanotoxicology testing is provided in Table 8.1. This table, although not intended to be comprehensive, includes examples of standards and measurement method development programs at various National Measurement Institutes (NMIs), in collaboration with the NMIs, within other government agencies, and within other documentary Standards Development Organizations.

One of the important benefits of documentary standards, particularly with nanotoxicology, is an agreed-upon and standardized approach to testing. With a lack of standards for the testing of conventional chemicals, history has clearly demonstrated that many types of results can be obtained even when a similar method is used. In toxicity testing, the variability in results may be caused by a wide range of factors including the state of the cell culture, or even the proper identification of the cell line [15, 16]. With nanomaterials, this is even more complicated as we superimpose a plethora of other variables including the physico-chemical properties of the nanomaterial itself that can lead to: (1) difficulties in preparing a reproducible sample for toxicology testing; (2) changes in the sample in different phases of the industrial pipeline and throughout its lifecycle; and (3) changes in the sample after exposure to relevant media (e.g., protein adsorption, agglomeration, dissolution, and others).

Ultimately, the development of standard methods and practices will improve the quality of the data and simultaneously improve the confidence in science regarding the safety of nano-products for the worker and the public. Additionally, the implementation of standard methods and practices will provide the scientific basis to make decisions about risk and exposure to nanomaterials. It is important to note, however, that the standards community needs to work diligently and quickly to promote standard methods and practices so that scientific data can be generated ahead of momentum to regulate away from nano-based products due to unsubstantiated public fear and concern.

Table 8.1 Examples of national efforts towards developing standards for toxicity testing of nanomaterials in National Measurement Institutes (NMIs) and Standards Developing Organizations (SDOs)

Country	NMI	SDO	Other	Some national efforts with emphasis on nanotoxicology
Argentina		X		• Working Group for "the development of standards in the areas of health, safety and environmental aspects of nanotechnologies" Methodologies and data quality analysis for risk assessment http://www.iram.org.ar
			X	El Centro Científico Tecnologico, Consejo Nacional de Investigaciones Científicas y Técnicas (CONICET) – Mendoza • Developing assays for: – Acute toxicity of nanostructured alumina in mammals – Distribution of nanostructured alumina in mammals – Toxicity of nanostructured alumina in vertebrates http://www.cricyt.edu.ar
Australia			X	National Industrial Chemicals Notification and Assessment Scheme (NICNAS) • Active role in international activities to develop best practice testing protocols and risk assessment methodologies http://www.nicnas.gov.au
Brazil		X		Committee for the special study of nanotechnology (CEE-89) • Developing practices of health, safety and environment with a scientific basis http://www.abnt.org.br
Canada	X			National Institute for Nanotechnology, Alberta • Developing "methods to assess the risk of nanomaterials," including developing *in vitro* assays, *in vivo* studies, and tests for systemic responses http://www.nrc-cnrc.gc.ca
China			X	Ministry of Science and Technology • Supports standardization activities in nanotechnology including health, safety and environment Bio-Environmental Health Sciences of Nanoscale Materials Laboratory, National Center for Nanoscience and Technology of the Chinese Academy of Sciences • Studies the nanotoxicology of manufactured nanomaterials • Includes "Innovative methodology for nanotoxicological studies"

Finland			X	Finnish Institute of Occupational Health • Proteomics study of the health effects of engineered nanoparticles • European NANOSH Conference: discussed global safety issues surrounding nanoparticles and nanotechnologies, in occupational safety and health in particular; included genotoxicity, pulmonary inflammation, and microcirculation effects of nanoparticles http://www.ttl.fi/Internet/English/Research/Research+database+TAVI/naytaProjekti?id=327982&type=Research%20project
Iran		X		Iran Nanotechnology Standardization Committee • Supporting the development of national standards for two methods for testing the toxicology of nanosilver: – Test method for evaluation of irritation and corrosion in response to nanosilver particles in an animal model for cosmetics applications – Evaluation of interaction of nanosilver with protein molecules for assessing nanosilver toxicity http://www.en.nano.ir/index.php/main/page/16
Japan	X			National Institute of Advanced Industrial Science and Technology (AIST) sponsoring several projects at the Ministry of Economy Trade and Industry (METI) since 2005: • Standardization of Nanoparticle Risk Evaluation Method with goals to: – Develop methods for the characterization of nanoparticles – Develop methods for assessment of health impact and safety of nanoparticles – Develop systems for data collection and data standardization • Risk assessment of manufactured nanomaterials http://www.aist.go.jp
			X	National Institute of Health Sciences (NIHS) • Development of assessment methods for health impact of nanomaterials
Korea		X		Korean Agency for Technology and Standards (KATS) • Workshop related to "NT safety assessment technology, research results of nanocosmetics' safety regarding the human body, and related standards development activities" http://www.ats.go.kr

(continued)

Table 8.1 (continued)

Country	NMI	SDO	Other	Some national efforts with emphasis on nanotoxicology
Singapore		X		Standards, Productivity and Innovation Board (*SPRING Singapore*), a statutory board under the Ministry of Trade and Industry of Singapore, has partnered the Agency for Science, Technology and Research (A* STAR) to lead standardization efforts in nanotechnology • Goal "to ensure the successful transition of nano-products from the laboratory to the market and address concerns over the impact of nanotechnology on health, safety and the environment" http://www.standards.org.sg/files/Vol15no1art3.htm
South Africa			X	Council for Scientific and Industrial Research (CSIR) • Research platform for nanotechnology health, safety and the environment (Nano-HSE) to "support focused research to address prioritized knowledge gaps such as nanomaterial characterization and dosimetry, environmental effects and exposure assessment, epidemiology/toxicology in mammalian systems, and risk assessment and risk management" http://www.csir.co.za
Spain	X			Supreme Council of Metrology (Centro Espanol de Metrologia, CEM) • Nano-materials: research towards traceability measures of toxicity, shape, size, distribution, chemical characterization of nano-particles such as combustion products or nano-tubes http://www.cem.es
Switzerland			X	Swiss Federal Laboratories for Materials Science and Technology (EMPA) • Working on developing standardized, validated protocols for the determination of nanospecific toxicities, specifically viability, inflammation, genotoxicity, oxidative stress http://www.empa.ch
Taiwan			X	Taiwan Nanotechnology Standard Council (TNSC), under the steering of National Nanotechnology Program and Bureau of Standards, Metrology & Inspection (MOEA) • Working group on "Health, Safety and Environmental issues of Nanotechnologies" http://www.itri.org.tw

UK		X		British Standards Institute (BSI) • National committee NTI/1 on "Nanotechnologies" http://www.bsigroup.com
			X	Safety of Nano-Materials Interdisciplinary Research Centre (SnIRC) • "Develop internationally agreed *in vivo* and *in vitro* protocols and models for investigating the routes of exposure, bioaccumulation and toxicology of nanoparticles in humans and non-human organisms" http://www.snirc.org
USA			X	National Nanotechnology Initiative (NNI) • Creates a framework for a comprehensive nanotechnology R&D program for the USA • Consists of the individual and cooperative nanotechnology-related activities of 25 US Federal agencies with a range of research and regulatory roles and responsibilities, including organizations with primary standards activities http://www.nano.gov
	X			National Institute of Standards and Technology (NIST) • Initiative for "Environment, Health and Safety Measurements & Standards" • "Develop detection and measurement methods for quantifying the number and nature of nanoparticles with EHS impact in biological and environmental samples" http://www.nist.gov/public_affairs/factsheet/environment2009.html
		X		American National Standards Institute (ANSI) Nanotechnology Steering Panel • Holds secretariat of ISO TC 229 Working Group 3 on Health, Safety and Environment http://www.ansi.org/standards_activities/standards_boards_panels/nsp/overview.aspx

This chapter examines the progress of documentary standards development related to nanotoxicology. We review the types of entities developing standards with a focus on the current status of standards development related to nanomaterial toxicity testing in these organizations. A second topic we review is the need for the validation of testing methods as applied to nanomaterials. Under this second topic, we highlight a few of the scientific efforts around the world that are aimed at developing and evaluating nanotoxicity testing methods, and whose efforts can be used to support the validity of international documentary standards and guidelines for nanotoxicity testing.

8.2 International Efforts Related to Nanotoxicology

In the international arena, several influential bodies have emerged as leaders in the area of standards and testing protocols for nanotoxicology. These include the Organization for Economic Cooperation and Development (OECD), the International Organization for Standardization (ISO), and ASTM International.

8.2.1 OECD

The OECD plays a pivotal role in coordinating national activities within the area of nanotechnology. The activity is centered around two Working Parties: Working Party on Nanotechnology, established in March 2007 "to promote international co-operation that facilitates research, development, and responsible commercialization of nanotechnology"; and Working Party on Manufactured Nanomaterials (WPMN), established in September 2006 "to promote international co-operation in human health and environmental safety related aspects of manufactured nanomaterials (MNs), in order to assist in the development of rigorous safety evaluation of nanomaterials."

There are 30 OECD member countries represented in the two working parties along with the European Commission, non-members (Brazil, China, Singapore, Thailand, Russia), ISO, World Health Organization (WHO), United Nations Environment Program (UNEP), and other relevant stakeholders. Most pertinent to this chapter are the activities of the WPMN with its focus on human health and environmental safety. The aim and objectives of OECD-WPMN are reported in *Manufactured Nanomaterials: Work Programme 2006–2008* [17], and the work is organized into eight Steering Groups shown in Table 8.2.

A fundamental role is currently given to the SG3 Sponsorship Programme to improve existing data and knowledge of the human health and environmental safety implications of manufactured nanomaterials that have potential for wide use and dissemination. The 14 identified materials are: fullerenes, single-walled carbon

Table 8.2 OECD-WPMN steering groups

SG1	Database on Human Health and Environmental Safety Research: Database with research project launched in March 2009
SG2	Research Strategy(ies) on Human Health and Environmental Safety Research: Review of current research programmes
SG3	Testing a Representative Set of Manufactured Nanomaterials (MN): Sponsorship programme for the testing of 14 materials for 61 endpoints
SG4	Manufactured Nanomaterials and Test Guidelines: Development of guidance on sample preparation and dosimetry for the testing of manufactured nanomaterials
SG5	Co-operation on Voluntary Schemes and Regulatory Programmes: Analysis of national information gathering programmes and regulatory frameworks
SG6	Co-operation on Risk Assessment: Review of existing risk assessment schemes and their relevance to nanomaterials
SG7	The Role of Alternative Methods in Nanotoxicology: Reviewing alternative test methods which will avoid animal tests and which will be applicable to manufactured nanomaterials
SG8	Exposure Measurement and Exposure Mitigation: Development of recommendations on measurement techniques and sampling protocols for inhalation and dermal exposures in the workplace

nanotubes, multi-walled carbon nanotubes, silver nanoparticles, iron nanoparticles, carbon black, titanium dioxide, aluminum oxide, cerium oxide, zinc oxide, silicon dioxide, polystyrene, dendrimers, and nanoclays. As noted:

> The list of endpoints is a set to take into account when testing specific MNs for human health and environmental safety within phase one of the Testing Programme. Addressing this set should ensure consistency between the various tests to be carried out on specific nanomaterials. It should also lead to the development of dossiers for each nanomaterial describing basic characterization, fate, ecotoxicity and mammalian toxicity information [18].

The list of 61 identified endpoints that will provide a base set of data with which to evaluate risk include: Nanomaterial Information/Identification (9 endpoints); Physico-Chemical Properties (16 endpoints); Environmental Fate (14 endpoints); Environmental Toxicology (5 endpoints); Mammalian Toxicology (8 endpoints); Material Safety (3 endpoints). The initial set of specific endpoints is listed in Table 8.3. It is also noted that this list of endpoints could be refined as the results of testing are assessed.

Table 8.3 OECD SG3, safety testing of a representative set of manufactured nanomaterials: endpoints by category that are to be reported during the assessment of 14 manufactured nanomaterials

Category	Endpoint
Nanomaterial information/ identification	□ Nanomaterial name (from list) □ CAS number □ Structural formula/molecular structure □ Composition of nanomaterial being tested (including degree of purity, known impurities or additives) □ Basic morphology □ Description of surface chemistry (e.g., coating or modification) □ Major commercial uses □ Known catalytic activity □ Method of production (e.g., precipitation, gas phase)
Physical-chemical properties and material characterization	□ Agglomeration/aggregation □ Water solubility □ Crystalline phase □ Dustiness □ Crystallite size □ Representative TEM picture(s) □ Particle size distribution □ Specific surface area □ Zeta potential (surface charge) □ Surface chemistry (where appropriate) □ Photocatalytic activity □ Pour density □ Porosity □ Octanol-water partition coefficient, where relevant □ Redox potential □ Radical formation potential □ Other relevant information (where available)
Environmental fate	□ Dispersion stability in water □ Biotic degradability □ Ready biodegradability □ Simulation testing on ultimate degradation in surface water □ Soil simulation testing □ Sediment simulation testing □ Sewage treatment simulation testing □ Identification of degradation product(s) □ Further testing of degradation product(s) as required □ Abiotic degradability and fate □ Hydrolysis, for surface modified nanomaterials □ Adsorption-desorption □ Adsorption to soil or sediment □ Bioaccumulation potential □ Other relevant information (when available)

(continued)

Table 8.3 (continued)

Category	Endpoint
Environmental toxicology	☐ Effects on pelagic species (short term/long term) ☐ Effects on sediment species (short term/long term) ☐ Effects on soil species (short term/long term) ☐ Effects on terrestrial species ☐ Effects on microorganisms ☐ Other relevant information (when available)
Mammalian toxicology	Pharmacokinetics (ADME) ☐ Acute toxicity ☐ Repeated dose toxicity If available: ☐ Chronic toxicity ☐ Reproductive toxicity ☐ Developmental toxicity ☐ Genetic toxicity ☐ Experience with human exposure ☐ Other relevant test data
Material safety	Where available: ☐ Flammability ☐ Explosivity ☐ Incompatibility

Current additional activities in the WPMN related to toxicological testing protocols are in SG4 and SG7. A guidance document was recently completed in SG4 describing issues related to sample preparation methods and dosimetry of nanomaterials for toxicological testing [19]. SG7 is in the process of preparing a guidance document describing developments in the toxicology field toward "animal-free" testing strategies as they pertain to nanotoxicology.

In general, the OECD WPMN activities provide key input into standards development activities within ISO and other organizations as they seek to: attain consensus from experts in the field regarding nanotoxicology testing; coordinate international testing activities; identify good practice in the nanotoxicology testing arena; and provide an indication of appropriate standards needed for the implementation of existing regulation and/or the development of new regulatory regimes for nanotechnologies.

8.2.2 ISO Technical Committee on Nanotechnologies

The ISO Technical Committee on Nanotechnologies (TC 229) was created in 2005 and is a robust and active committee with 32 participating countries, 11 observing countries, and seven organizations in liaison including OECD, Versailles Project on Advanced Materials and Standards (VAMAS), Asia NanoForum (ANF HQ), Bureau International des Poids et Mesures (BIPM), European Environmental

Citizens Organisation for Standardization (ECOSS), European Union (EU), and the Institute for Reference Materials and Measurements (IRMM). It is currently structured into four working groups related to terminology, measurement and characterization, environmental health and safety, and materials specifications.

The Working Group on Health, Safety and Environmental Aspects of Nanotechnologies (Working Group 3) was one of the working groups that defined the original structure and serves as the home in TC 229 for documentary standards related to nanotoxicology. The roadmap for the working group lays out a plan for the development of a suite of standards in the following areas: (1) Standard Methods for Controlling Occupational Exposures to Nanomaterials; (2) Standard Methods for Determining Relative Toxicity/Hazard Potential of Nanomaterials; (3) Standard Methods for Toxicological Screening of Nanomaterials; (4) Standard Methods for Environmentally Sound Use of Nanomaterials; (5) Standard Methods for Ensuring Product Safety of Nanomaterial Products; and (6) Standard Methods to Support OECD Working Party on Manufactured Nanomaterials. The first guidance document that emerged from this group was related to nanomaterial occupational health and safety. A list of current work items in the group is provided in Table 8.4, and while only one of those in the pipeline specifically addresses methods for assessing the toxicological properties of nanomaterials, several others are still pertinent to this chapter and are discussed here. For the most current list of all standards under development in ISO TC 229, see [20].

By now, it is well known that the misinterpretation of data related to the toxicity of nanomaterials has been linked to poor or non-existent characterization of nanomaterials and/or their chemical or biological contaminants prior to toxicity testing.

Table 8.4 List of work items in ISO TC229 Working Group 3

Nanotechnologies – Endotoxin test on nanomaterial samples for *in vitro* systems – Limulus amebocyte lysate (LAL) test (ISO/FDIS 29701)
Nanotechnologies – Generation of metal nanoparticles for inhalation toxicity testing using the evaporation/condensation method (ISO/DIS 10801)
Nanotechnologies – Characterization of nanoparticles in inhalation exposure chambers for inhalation toxicity testing (ISO/DIS 10808)
Nanotechnologies – Guidance on physico-chemical characterization for manufactured nano-objects submitted for toxicological testing (ISO/PDTR 13014)
Nanotechnologies – Guidance on safe handling and disposal of manufactured nanomaterials (ISO/AWI TS 12901-1)
Nanotechnologies – Nanomaterial Risk Evaluation Framework (ISO/AWI TR 13121)
Nanotechnologies – Guidelines for occupational risk management applied to engineered nanomaterials based on a "control banding approach" (ISO/NP TS 12901-2)
Nanomaterials – Preparation of Material Safety Data Sheet (MSDS) (ISO/NP TR 13329)
Nanotechnologies – Compilation and Description of Toxicological and Ecotoxicological Screening Methods for Engineered and Manufactured Nanomaterials
Nanotechnologies – Compilation and Description of Sample Preparation and Dosing Methods for Engineered and Manufactured Nanomaterials
Nanotechnologies – Surface characterization of gold nanoparticles for the identification of bound molecules before and after cytotoxicity test: FT-IR method

Note that all documents are in process and titles may be subject to change

Additionally, because of the lack of physico-chemical characterization data, there is little to underpin our understanding of structure-activity relationships and predictive toxicology based on these relationships. Therefore, in recent years there has been a strong push by the community of researchers, regulators, and others toward a minimum set of characterization data that would improve our understanding of the link between nanomaterials and their potentially toxic effects. In support of this trend, two documents in the ISO pipeline aim to provide guidance on physico-chemical characterization and screening of contamination broadly applicable to *in vitro* and *in vivo* nanotoxicity testing. These are: Guidance on Physico-Chemical Characterization for Manufactured Nano-objects Submitted for Toxicological Testing; and Endotoxin Test on Nanomaterial Samples for *In Vitro* Systems – Limulus Amebocyte Lysate (LAL) test. The technical report, Guidance on Physico-Chemical Characterization for Manufactured Nano-objects Submitted for Toxicological Testing, is in draft form and due to be completed in early 2011. This document is anxiously awaited by the community as it provides a list, developed in concert with OECD and other national and international entities, of nanomaterial parameters that are recommended to be measured prior to toxicological screening. The current list of parameters includes surface properties (such as surface chemistry, surface area, and surface charge) and bulk properties (such as composition, size, shape, agglomeration/aggregation, solubility, and dispersibility), and was developed based on the current state of our understanding of the toxicological significance of various nanomaterial characteristics. The final document will include the complete list of these recommended parameters, discussion of the toxicological relevance of each parameter, and a list of potential methods available for measuring each parameter. It is envisioned that adoption and realization of this list by the community will have a profound impact on our ability to correctly assess nanomaterial toxicity and to correlate nanomaterial property with toxicological effect. Some have suggested that a minimum set of physico-chemical parameters be required for publication of nanotoxicity studies; however this has sparked lively debate and controversy in several open forums (i.e., Society of Toxicology Nanotoxicology specialty section; http://www.toxicology.org/isot/ss/nano/news.asp).

Prior to testing nanomaterials for toxicity, another critical step is to characterize the sample for the presence of biological contaminants, particularly endotoxins, on the nanomaterials. Endotoxins are mostly lipopolysaccharide (LPS) components of gram negative bacterial membranes. Lipopolysaccharides are released into the environment when the bacterial cells are lysed, but also bacteria shed these components as a part of their normal life cycle. This shedding process is a major source of endotoxin contamination. Endotoxins are heat stable and therefore do not degrade under a variety of normal environmental and laboratory conditions. They induce a potent toxic response in *in vitro* assays involving macrophages and other mammalian cells, and also induce a positive toxic response in *in vivo* mammalian systems. It is important to ascertain contamination of any type of sample by endotoxins prior to toxicological assessment; however, the problem is exacerbated with nanomaterials where the high surface area of nanomaterials in a sample can lead to accumulation of large amounts of chemical or biological contaminant.

A standard completed in 2010, Endotoxin Test on Nanomaterial Samples for *In Vitro* Systems – Limulus amebocyte lysate (LAL) test, describes a method for determining the presence of endotoxin on nanomaterials to prevent misinterpretation of the toxicity testing results.

Applicable to *in vivo* toxicity testing of nanomaterials, the Working Group has two documents that were recently completed: Generation of Metal Nanoparticles for Inhalation Toxicity Testing using the Evaporation/Condensation Method; and Characterization of Nanoparticles in Inhalation Exposure Chambers for Inhalation Toxicity Testing. The second item emphasizes the importance of good nanomaterial characterization under the proper conditions of, and at the point of, exposure in the animal chamber.

Two new work item proposals balloted in early 2010 and initiated later in the same year within the WG3 are: (1) Compilation and Description of Sample Preparation and Dosing Methods for Engineered and Manufactured Nanomaterials; and (2) Compilation and Description of Toxicological and Ecotoxicological Screening Methods for Engineered and Manufactured Nanomaterials. The initial scope and intent of the first document, Compilation and Description of Sample Preparation and Dosing Methods for Engineered and Manufactured Nanomaterials, is to provide a compilation and description of sample preparation and dosing methods, and to discuss physico-chemical properties, media, and other considerations as they pertain to or impact selection of an appropriate sample preparation method. As the OECD WPMN has completed guidance that discusses considerations and caveats associated with sample preparation for nanotoxicology testing, it is intended that the ISO guidance document will coordinate closely with OECD on this project. The OECD document will be a critical starting reference that may be enhanced and updated based on knowledge that develops over the course of the 2 years that the ISO document is under development. Additionally, since the intent of the ISO document is to produce guidance that can frame its standards in this area, the ISO document will add lists of current methods; and identify potential methods that can be developed into future standards related to sample preparation for nanomaterials in *in vitro* and *in vivo* screening.

The initial scope and intent of the second document, Compilation and Description of Toxicological and Ecotoxicological Screening Methods for Engineered and Manufactured Nanomaterials is to provide a compilation and description of toxicological screening methods that have been applied to the testing of manufactured nanomaterials. There are national efforts in several countries that are evaluating toxicological screening methods and it is intended that these national efforts will be cited in the guidance document. As with the previous document, one main purpose will be to identify future standard methods related to assays for *in vitro* and *in vivo* screening of nanomaterials.

The Nanomaterial Risk Evaluation Framework guidance document contains some text describing the use of tiered testing approaches for toxicology testing; however, the main focus of that document is not nanotoxicology, but to create an acceptable framework that manufacturers and users could apply to ensure sound risk management of their nanotechnology products across the product lifecycle.

Other activities within ISO that deal with toxicity testing, but are not specifically related to nanomaterials, reside primarily in ISO TC 194. ISO TC 194 on Biological Evaluation of Medical Devices has an extensive portfolio related to *in vitro* and *in vivo* toxicity testing of medical devices as well as documents related to physico-chemical characterization, sterilization, and sample preparation. ISO TC 194 has liaison status with TC 229 and therefore can contribute to the development of standards in nanotoxicology testing.

8.2.3 ASTM International

In 2005, ASTM International also established a Technical Committee on Nanotechnology, E56, to develop standards and guidances related to nanotechnology and nanomaterials.

Activities related to the development of standards and guidances for toxicological testing are primarily under the purview of E56.02, Characterization: Physical, Chemical, and Toxicological Properties. In 2008, three standards were published related to *in vitro* toxicological testing of nanomaterials as shown in Table 8.5. E56.02 currently has no proposed new standards listed that fall in the category of standards for toxicological testing.

E2524-08 describes a method for determining the hemolytic effect of nanomaterials by measuring hemoglobin release from damaged red blood cells. This method employs a simple spectrophotometric assay that measures methemoglobin concentration following oxidation of released hemoglobin (and its derivatives) in the presence of ferricyanide in alkali solution. A similar method was described in ASTM standard F756 for assessing the hemolytic properties of other materials that are not specifically nanoscale.

E2525-08 is a method for determining the effect of nanomaterials on the formation of granulocyte and macrophage colonies from bone marrow stem cells cultured in physiological solution. The assay result is the total number of colony forming units (granulocyte, macrophage and granulocyte-macrophage colonies) measured after 12 days in culture. The effect of the nanomaterials can be neutral, stimulatory, or inhibitory on these cells associated with the immune system, and the result of the

Table 8.5 List of published standards from ASTM E56 related to toxicological testing

Reference number	Title
E2524-08	Standard Test Method for Analysis of Hemolytic Properties of Nanoparticles
E2525-08	Standard Test Method for Evaluation of the Effect of Nanoparticulate Materials on the Formation of Mouse Granulocyte Macrophage Colonies
E2526-08	Standard Test Method for Evaluation of Cytotoxicity of Nanoparticulate Materials in Porcine Kidney Cells and Human Hepatocarcinoma Cells

assay is used to infer potential health effects associated with interference in immune cell differentiation and growth.

E2526-08 describes a method for measuring nanoparticle cytotoxicity with two cells types (HEP-G2 and LLC-PK1) that may indicate the effect on two potential target organs: liver and kidney. Two methods for assessing cytotoxicity in these cell lines are described: (1) the MTT assay that demonstrates a decrease in cell viability in the presence of nanoparticles as measured by a decrease in the activity of metabolic enzymes to reduce the MTT [3-(4,5-di**M**ethyl**T**hiazol-2-yl)-2,5-diphenyl**T**etrazolium bromide] reagent; and (2) the LDH leakage assay that demonstrates membrane damage by measuring release of an enzyme found in plants, animals and humans called **L**actate **D**e**H**ydrogenase (LDH) from cells in the presence of nanoparticles in a colorimetric assay using the tetrazolium salt, 2-*p*-iodophenyl-3-*p*-nitrophenyl-5-phenyl tetrazolium chloride, as the substrate.

ASTM also supports its standards through the Interlaboratory Study (ILS) Program: a testing program designed to provide determination of repeatability and reproducibility of the methods described in its standards through interlaboratory comparisons. Participation in the ASTM interlaboratory studies is voluntary and publication of ASTM standards is not contingent upon the results of any interlaboratory study. In fact, publication of the standard can precede the completion of an interlaboratory study, but with the desired outcome being incorporation of this data into later revisions of the standard. In 2009, E56 had three interlaboratory studies registered with two of these pertaining to standards for *in vitro* toxicological testing: (1) New Standard Practice for Analysis of Hemolytic Properties of Nanoparticles; and (2) New Standard Practice for Evaluation of Cytotoxicity of Nanoparticulate Materials on Porcine Kidney Cells. These studies will be discussed further in Sect. 8.3 of this chapter.

8.3 Need for Validation in Toxicity Testing and Assay Validation Efforts

8.3.1 Need for Validation in Toxicity Testing

Validation is the process by which the assay is determined to be reproducible (i.e., result is the same when repeated), reliable (i.e., result is accurate), and robust (i.e., result is most often correct). Although there are sets of assays that are regularly used in toxicology testing to assess cytotoxicity, genotoxicity, effect on cells of the immune system, etc., many of these have not been validated for use in the testing of nanomaterials. There are already indicators that some specific nanomaterials interfere with the results of certain common tests and therefore these assays are not valid, e.g., carbon nanoparticles interfering with results of the MTT test and the LDH test described above. Kroll et al. [21] provide a detailed list with references that demonstrate interferences of nanoparticles with these and other common toxicity assays through a number of different mechanisms. It is important to note that although an assay may be validated and therefore suitable for use with one type of nanomaterial, it may not be suitable for use with other types of nanomaterials and therefore each assay should be validated with each class or type of nanomaterial.

Validation of toxicity testing methods for nanomaterials is currently a focus of many scientific efforts. Assay validation for the development of international standards is best accomplished across multiple laboratories using identical samples of nanomaterials and tested using prescriptive methods. Once an assay is determined to be suitable for use with nanomaterials without interference, each new individual laboratory is responsible for performing internal validation to verify that it has the capability to carry out that assay successfully. Some efforts to develop, describe and validate toxicological screening methods for use with nanomaterials are described below. It is anticipated that these method validation efforts will promote the development and dissemination of toxicity testing protocols and standards for use with nanomaterials.

8.3.2 Efforts to Support Validation of Toxicological Testing and the Development of Standards

8.3.2.1 NCI-NCL

Supporting the US National Cancer Institute's (NCI) Challenge Goal of eliminating death and suffering from cancer, the NCI is harnessing the power of nanotechnology to fundamentally change the way to detect, diagnose, treat, and prevent cancer. Development of technologies to control the physical, chemical, and biological properties of nanoparticles enables new ways to engineer and use these materials in cancer prevention, diagnostics, and therapy [22]. In 2005, the Nanotechnology Characterization Laboratory (NCL) was established within NCI and initially in partnership with FDA and NIST to provide critical infrastructure support to this emerging field. The intent of the NCL is to accelerate the transition of basic nanobiotechnology research into cancer treatment clinical applications. Among the NCL priority research areas is to establish and standardize an analytical test cascade for the pre-clinical toxicology, pharmacology, and efficacy of nanoparticles. NCL characterizes nanomaterials from academia, government, and industry for their physical properties, *in vitro* biological properties, and *in vivo* compatibility through the use of animal models [23].

Nanomaterial physical and chemical characterization protocols are being developed in cooperation with NIST focusing on the following properties: size and size distribution, topology, molecular weight, aggregation, purity, chemical composition, surface characteristics, functionality, zeta potential, stability, and solubility, some of which have been linked to nanomaterial distribution and fate *in vivo*. One standard NIST method for this set of parameters, Measuring the Size of Nanoparticles in Aqueous Media Using Batch-Mode Dynamic Light Scattering, NIST-NCL PCC-1, was published in 2007 [24] with research on other methods currently under development.

In vitro nanomaterial testing protocols conform to existing US Food and Drug Administration (FDA) requirements of toxicity or biocompatibility studies for Investigational New Drug (IND) or Investigational Device Exemption (IDE) applications. Established cell and molecular biology methods are used to monitor nanoparticle binding, pharmacology, blood contact properties, interaction with cellular-level components, and the nanomaterial's therapeutic and/or diagnostic functionality. *In vitro* models are used as a gross approximation of nanomaterial absorption, distribution, metabolism, excretion, and toxicity properties. The NCL currently maintains 24 publicly available methods for *in vitro* toxicity testing that are accessible from their website [25]. It is important to note that many of these assays have been tested for use with more than 100 nanomaterials that were designed and intended for medical use.

NCL's *in vivo* characterization is based on protocols used in the U.S. to characterize drugs and devices. These measurements will describe the nanoparticle's absorption, pharmacokinetics, serum half-life, protein binding, tissue distribution/accumulation, enzyme induction or inhibition, metabolites, and excretion pattern. Two primary goals are set: (1) to identify nanoparticle doses causing no adverse effect; and (2) to identify doses causing life-threatening toxicity. The NCL does not currently maintain any publicly available methods for *in vivo* toxicity testing. The NCL toxicity testing methods for determining a nanomaterial's "no adverse effect levels" may be invaluable in accelerating the development of new toxicity testing methodologies for nanomaterials.

Although the validation of their methods was done internally, the NCL is actively engaged in relevant international efforts related to standardization, testing, and validation of laboratory methods for nanotoxicology.

8.3.2.2 International Alliance for NanoEHS Harmonization

In September 2008, a group of scientists involved in nanotechnology research from Europe, Japan, and the United States formed a peer group called The International Alliance for NanoEHS Harmonization (IANH) [26]. IANH is dedicated to establishing reproducible approaches for the study of nanoparticle interactions with living organisms, with an emphasis on supporting world-wide efforts on nanosafety. The Alliance is currently chaired by Dr. Kenneth Dawson of the Center for BioNano Interactions at University College Dublin. The primary goal of the Alliance is to create a trustworthy co-operation between scientists where results are simultaneously and carefully checked using round robin studies. IANH is committed to pursuing the understanding of bio-nano interactions at the scientific level and supports efforts of organizations such as OECD, ISO, and ASTM by upstream identification of sources of irreproducibility and by offering potential solutions. Test materials for the first round robin studies include TiO_2, CeO_2, ZnO, Ag, Au,

MWCNTs and polystyrene nanoparticles. A single batch of several cell lines (A549, BEAS-2B, RAW264.7) was sourced and initial characterization of these cell lines was completed by six alliance partners. Nanoparticle dispersion protocols in buffer and cell culture media are in development and will incorporate full characterization of size and size distribution, zeta potential, redox potential, and radical formation potential in both media. Nanoparticle preparation and material shipping protocols consist of detailed procedures for elution, cleaning, dispersion, and measurement of contaminant levels. *In vitro* effects are being tested using assays for MTT, LDH, cell reactive oxygen species (ROS), cytokine induction, and genotoxicity (COMET assay) [27]. The health of the cells will be assessed by monitoring their growth rate and the uptake rate of detectable nanoparticles (Au and fluorescent polystyrene). To evaluate reproducibility, the biological and toxicological responses will be measured by each partner lab and the results will be posted in the IANH web site for comparison to other results. Following detailed protocol review against the procedure of the lead lab, the validation lab will repeat the experiments and the results will be checked again. *In vivo* characterization will include the biokinetics of the nanoparticles and the toxicology in the organs of the animals, such as "common" rodents and environmentally sensitive aquatic species (Daphnia, Zebra Fish and *C. elegans*). These results will be compared to the *in vitro* data as a function of exposure dose, surface area, and dose rate. Once these protocols are validated for reproducibility, they will be presented to standards organizations for consideration.

8.3.2.3 ASTM International's Interlaboratory Studies

As noted previously, ASTM E56 is conducting two interlaboratory studies ("round robins") that are designed to assess precision and bias of their nanotoxicity assay standards: (1) ILS201 – to test New Standard Practice for Analysis of Hemolytic Properties of Nanoparticles; and (2) ILS202 – to test New Standard Practice for Evaluation of Cytotoxicity of Nanoparticulate Materials on Porcine Kidney Cells. Although there is no published report available from these studies to date, a published comment is available on the web [28]. A workshop was held at NIST in October 2008 to share results of these interlaboratory comparisons with the community in an open forum.

There were nine participating laboratories in ILS201 and six participating laboratories in ILS 202, and these participants were supplied with the following samples for testing: (1) NIST RM colloidal gold, 30 nm nominal diameter; (2) NIST RM colloidal gold, 60 nm; (3) cationic dendrimer (positive control); (4) neutral dendrimer (negative control). In the hemolysis study (ILS201), the majority of laboratories were unable to complete the entire study to supply data on all samples; however, data that were complete showed that the assay could be performed successfully. It was noted that difficulties with sample preparation most often led to failure of the assay. Data in the cytotoxicity study (ILS202) was also incomplete and it was noted that this assay required better (more toxic) positive controls. With both studies it was determined that participating laboratories should be required to perform training sets prior to participating in formal studies on nanotoxicology because of the special issues associated with preparing/handling these materials, emphasizing the attention to sample preparation in future standards. More studies are planned with the ultimate goal of providing supporting data on reproducibility for the nanotoxicity standards under development in ASTM. Lessons learned from these studies are also being shared with the community. This approach should more broadly feed the development of future standards and provide context about which standards are needed by exposing weaknesses and sources of uncertainty in the measurement process.

8.3.2.4 European Network on the Health and Environmental Impact of Nanomaterials and European Center for the Validation of Alternative Methods

The stated objective of the European network on health and environmental impact of nanomaterials [29] is to create a scientific basis to ensure the safe and responsible development of engineered nanoparticles and nanotechnology products, and to support regulatory measures and implementation of legislation in Europe. It aims to develop a structure for the critical evaluation of methods and protocols and is a 4-year project funded by the European Commission's Seventh Framework Programme (FP7). The network consists of 24 institute partners that are leading European research groups active in the fields of nanosafety, nanorisk assessment and nanotoxicology. By coordinating research efforts of scientists from across the EU countries, NanoImpactNet helps to harmonize methodologies and communicate results, initially across Europe, and later worldwide, with the goal of facilitating consensus building and identification of strategies to address knowledge gaps. NanoImpactNet provides an online space for sharing nanotoxicological protocols between members. It is intended therefore, that laboratories can easily compare their methods and subsequently develop common protocols and strategies for the testing of nanomaterials. Only protocols that are established within laboratories and published in peer-reviewed journals can be submitted. Selected NanoImpactNet members are given the opportunity to download the protocol, test it and comment on its advantages and disadvantages within the NanoImpactNet online community. These comments are then made available to all NanoImpactNet members for an open discussion. Following this discussion, the protocol can be upgraded to a protocol recommended by NanoImpactNet and made available to all on the website. Finally, the protocols are published in a format conforming to international standards and are provided to international bodies as a "Method recommended by NanoImpactNet."

The recent NanoImpactNet report entitled, "First Approaches to Standard Protocols and Reference Materials for the Assessment of Potential Hazards Associated with Nanomaterials," highlighted the urgent need for nanoparticle reference (test) materials and the need to share protocols and best practice.

The reference materials identified by this group as being useful for ecotoxicology/environmental studies include TiO_2, polystyrene nanoparticles labeled with fluorescent dyes, and silver. Nanomaterial physico-chemical characteristics recognized by this group as important for toxicological assessment and recommended for standardization include indicators of aggregation/agglomeration/dispersibility, size, solubility, surface area, charge and chemistry. Cytotoxicity, particle uptake, oxidative stress, immune response and genotoxicity are listed as the most relevant endpoints for nanoparticle *in vitro* testing. A broad range of *in vivo* testing topics was addressed at several recent workshops, organized by the network. They include available nanomaterial labeling and tracking techniques, adequacy of transgenic mouse models, novel *in vivo* approaches and potential endpoints. Moreover, the network is actively debating *ex vivo* approaches as possible alternatives to *in vivo* and *in vitro* testing in assessing nanotoxicity [30].

A central role coordinating the search and implementation of testing methods aiming at the replacement, reduction or refinement of the use of laboratory animals for experimental and other scientific purposes is played by European Center for the Validation of Alternative Methods (ECVAM). It became operational as a unit within the EU Joint Research Center in 1993 and is focused on the development and evaluation of *in vitro* methods and of computer modeling using structure-activity relationships for toxicological assessment. ECVAM promotes the development of alternative testing methods by funding and organizing workshops as well as a limited number of external studies on test development that fit ECVAM's work programme. In addition, ECVAM performs applied research on test development and mathematical models predicting toxicological endpoints [31]. In short, ECVAM mission is to support the EU policies in the field of consumer protection, environmental protection and animal protection by validating alternative methods for toxicology testing that provide similar or better basis for risk assessment and management as *in vivo* tests by promoting their development, application and acceptance by regulators. As of 2010, 12 alternative methods for chemicals/cosmetics have been endorsed by ECVAM including skin corrosivity, skin sensitization, phototoxicity, acute fish toxicity, myelotoxicity, mutagenicity and embryotoxicity. ECVAM has established a wide international network with OECD, similar organizations in the U.S. (ICCVAM), Japan (JaCVAM), and European Commission Environment, Enterprise, Health and Consumer Protection General Directorates.

8.3.2.5 NanoInteract

NanoInteract is funded under the European Commission's Sixth Framework Programme under the NMP theme. NanoInteract aims to coordinate the efforts of European, U.S. and Israeli partners to validate established chemical toxicity testing for application to nanoparticle toxicity testing, and to identify ways in which the presence of nanoparticles and their aggregates can impact these tests or their interpretation (http://www.nanointeract.net) [32].

Towards the development of a platform and toolkit for understanding interactions between nanoparticles and the living world, NanoInteract has the following goals:

1. To establish experimental protocols for every aspect of the study of nanoparticle interaction with cells, and several types of aquatic plants and organisms, ensuring complete reproducibility.
2. To understand effect of adsorbed protein on nanoparticle stability and nanoparticles on protein conformation and function, ultimately connecting this to biological impacts.
3. To connect cellular location of nanoparticles with intra- and inter-cellular processes disrupted.
4. To combine these results, along with the expertise from diverse disciplines, to point towards a "standard approach to nanotoxicology."

The first cross-institutional round-robin toxicology experiment was conducted using silica nanoparticles from two independent sources in a range of nominal sizes (10–400 nm) and different surface properties. The 3T3 cell line was tested for genotoxicity using COMET assay and no genotoxicity was observed in all participating laboratories. The concept of the biomolecular corona (protein and lipid surrounding the nanoparticle) was first outlined in the NanoInteract project and has gained considerable interest and support in the scientific community [33].

A standard technique and protocol has been developed to deduce the major components of the nanoparticle corona. It is anticipated that this could be developed as a standard characterization in conjunction with IRRM and NIST for their silica and gold standard material nanoparticles. Also, the project produced the first computational model of nanoparticle uptake fitted to experimental results.

8.4 Future

8.4.1 Where Standards Are Needed: Opportunities for the Future

A variety of weaknesses have been exposed in our ability to perform *in vitro* toxicity testing on nanomaterials to obtain the same result from laboratory to laboratory. This is readily observed by scanning the literature in this field, and by participating in discussions of results of interlaboratory intercomparison studies designed to directly assess comparability of these results. Some of these weaknesses pertain to the wide number of variables in any biological experiment; however, in nanotoxicology experiments, this is confounded by the fact that samples containing nanomaterials are often not well characterized or purified, and they are difficult to prepare as reproducible colloidal dispersions in the appropriate biological media. Even if the nanomaterials are well characterized and adequately dispersed in the biological media, there may be additional unpredicted interferences using traditional *in vitro* toxicity testing methods as discussed previously. The community has also not come to agreement on an appropriate set of positive and negative controls for nanotoxicology experiments that can be used to adequately underpin the measurement result. A description of some of these variables is shown in Fig. 8.1 emphasizing the difficulty in performing measurements to assess nanomaterial toxicity even with relatively simple *in vitro* systems aimed at understanding human health effects.

Studies that have been done that highlight the weaknesses of nanotoxicology experiments are useful in that they provide insight into the types of standards that would serve the community well over the next 3–5 years. At the most fundamental level, the community requires standard methods and reference materials to support fundamental physical and chemical characterization of nanomaterials. It is hoped that the good measurement of these nanomaterial characteristics will provide the core of all understanding of the relationship between the nanomaterial and its toxicity. Many of the methods that are being developed as standards for physical and chemical characterization are specific for a single type of nanomaterial and not for broad classes of nanomaterials. However, where scientifically feasible and appropriate, measurement methods that can be applied to many types of nanomaterials should be developed as these are desired by the community.

The difficulties associated with the preparation of nanomaterials in appropriate media to produce a well-dispersed sample make sample preparation a surprising

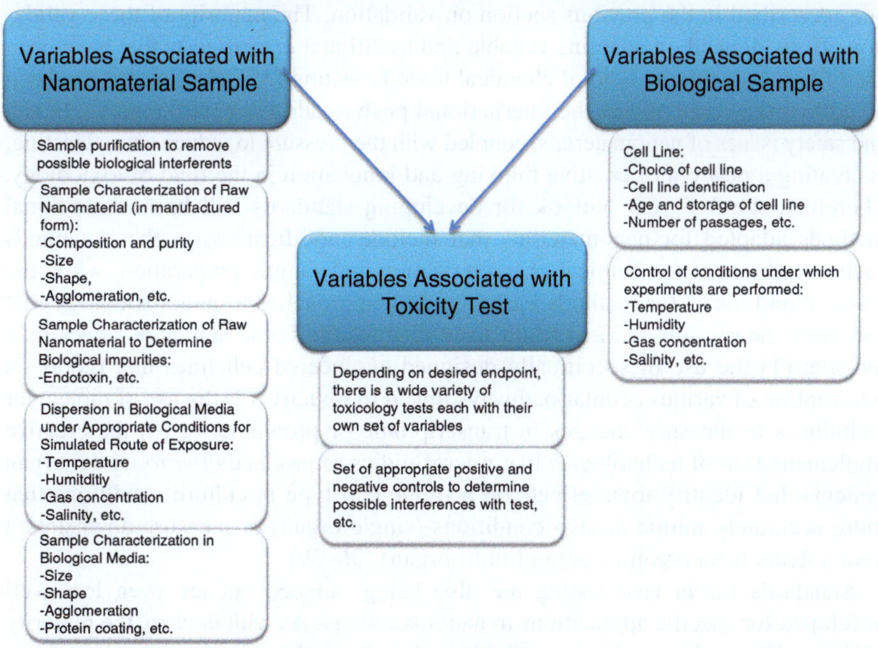

Fig. 8.1 Variables associated with *in vitro* toxicity testing

and difficult bottleneck in toxicity testing. For example, researchers have obtained varying agglomeration results depending on the dispersion method, such as bath sonication, probe sonication, and adding dispersing agents that include proteins, surfactants, and pyrophosphate [34]. Because the results vary depending on the dispersion method, overarching guidance documents discussing sample preparation for a variety of materials are in process in two organizations, OECD and ISO, as discussed previously. Standards that precisely prescribe sample preparation procedures for specific nanomaterials are also clearly desired by the community and should follow the development of these overarching documents. It is likely that many prescriptive methods for sample preparation will be embedded in the text of standard methods for toxicity testing and not developed as separate standards in many cases. Additionally, it is important to note that optimal sample preparation procedures may be linked to nanomaterial properties such as surface chemistry and surface charge; therefore, when a sample preparation procedure is not available, measurement of physico-chemical characteristics using standard methods may be used to guide the sample preparation approach.

With both *in vitro* and *in vivo* testing, standardized methods should include, or refer to, methods for sample characterization; methods for preparing materials for studies (e.g. aerosol generation, solubilization, and others); appropriate sampling methods; and guidance on relevant positive and negative reference materials and controls.

The first standard methods related to nanotoxicology will most likely be *in vitro* methods for toxicity testing of nanomaterials that will develop out of efforts like the ones described in the previous section on validation. The majority of these validation efforts depend on adapting reliable and traditional *in vitro* tests that have been used for decades to the field of chemical toxicity testing [35]. However, it has been noted by many experts that the international push to address environmental, health and safety issues of nanomaterials, coupled with the pressure to reduce animal testing, is creating a new era of creative thinking and innovation in the field of toxicology. Therefore, the 3–5 year outlook for developing standards will be for traditional methods adapted for nanomaterials that include modifications to these methods such as references to chemical characterization and sample preparation, while the >5-year outlook is for methods based on new approaches to nanotoxicology that can more clearly link *in vitro* result to *in vivo* result. These new approaches may include: (1) the use of specifically designed engineered cell lines that report on interruption of various cellular pathways linked to toxicity; (2) the use of molecular techniques to measure changes in transcriptome or proteome; or (3) the creative implementation of technologies like microfluidics to produce *in vitro* model organ systems that identify toxic effects on a single cell type in culture conditions that more accurately mimic *in vivo* conditions (single organ) or measure downstream toxic effects from organ-to-organ (multi-organ) [36–39].

Standards for *in vivo* testing are also being pursued but are even less well developed for specific applications in nanotoxicology. As with *in vitro*, the majority of these efforts rely on adapting reliable and traditional *in vivo* tests that have been used for decades to the field of chemical toxicity testing. There is, however, a word of caution, as OECD (and others) note:

> …Very little is known about the physiological responses to nanoparticles. Although some conventional toxicity and ecotoxicity tests have been shown to be useful in evaluating the hazards of nanoparticles, existing methodologies may require modification regarding hazard evaluation… [40]

The first standard methods and guidelines for *in vivo* testing in mammalian systems may include some of the following: pharmacokinetics (ADME), acute toxicity, repeated dose toxicity, irritation/corrosion, immunotoxicity/sensitization, chronic toxicity, reproductive toxicity, developmental toxicity, genetic toxicity, and experience with human exposures. Chronic toxicity is likely to include both cancer and non-cancer effects. Genetic toxicity could include a number of *in vivo* assays used today, including chromosome aberrations in bone marrow and Sister Chromatid Exchange assays.

One group that has been systematically addressing guidelines for toxicity testing (including both *in vitro* and *in vivo* measurements of potential health effects) is the OECD. Guidelines for the Testing of Chemicals (Test Guidelines) 4 are a collection of the most relevant internationally agreed testing methods used by government, industry and independent laboratories to assess the safety of chemical products. To date, OECD has published 118 test guidelines, which are organized in five sections: (1) Physical Chemical Properties; (2) Effects on Biotic Systems; (3) Degradation and Accumulation; (4) Health Effects; and (5) Other Test Guidelines. Relevant to this discussion, Sect. 4 current testing guidelines were reviewed for their adequacy for nanomaterials. In July 2009, OECD reported:

> The review of health effects related test guidelines (Sect. 4) concluded that, in general the OECD guidelines are applicable for investigating the health effects of nanomaterials with the important proviso that additional consideration needs to be given to the physico-chemical characteristics of the material tested, including such characteristics in the actual dosing solution. In some cases, there will be a need for further modification to the OECD guideline. This applies particularly to studies using the inhalation route and to toxicokinetic (ADME) studies. Finally, it is important to build upon current knowledge and practical solutions in relation to *in vitro* test approaches [41].

Given some of the questions raised by the OECD and others, are there issues that can be addressed in the short term that will have a great impact on progress in the field of nanotoxicology testing and standards? Related to the issue of dose, the foundation of toxicology is the measurement of dose-response. As noted in the introduction, since the time of Paracelsus in the fifteenth century, toxicology defined toxicants (and pharmacological agents) in dose and response. Dose has uniformly meant mass; specifically, units are mass per mass of the organism. Nanotechnology appears to be expanding the concept of dose. The current thought is that shape, surface area, surface chemistry and other physical and chemical parameters will be equally important as mass in the determination of dose. Coming to a unified agreement on the issue of dose will help to make comparisons across data sets from different laboratories more convincing.

Related to the issue of the physico-chemical characteristics of the material, it is important to consider whether the form of the material that is undergoing testing is consistent with the use of the nanomaterial in the consumer or workplace environment. Physical and chemical characterization should consider the form out of the bottle, the form administered, and possibly the form after it enters the body. Some have argued that raw materials used in manufacturing today (out of the bottle only) are useful test substances; however, there is increasing interest in understanding how multiple chemicals, such as those incorporated in a consumer product, could either mitigate or express toxicity. The increasing use of nanomaterials in products will likely increase the interest in testing specifically related to the product form.

Another area mentioned in the OECD report that will require reevaluation in nanotoxicology testing is the concept of exposure route. In current toxicology, exposure routes for environmental toxicants include dermal, ingestion, and inhalation. For pharmacological agents, common exposure routes also include intraperitoneal and intravenous exposures. Nanotechnology will likely provide additional routes as some agents might be small enough for unique exposure routes. For example, nano-sized manganese oxide was demonstrated to be absorbed via the olfactory neuron and transported to the brain of mice.

In summary, it is critical for the field of nanotoxicology testing to have standards based on widely accepted and scientifically proven methods and protocols. However, it is noted that there is still a great deal of research that needs to be done in all areas that will delay the near-term development of prescriptive standards in this area. The community must also come to some agreement on overarching issues related to dose, appropriate test materials, and exposure routes on which the development of future standard methods will rely.

8.4.2 Harmonization and the Role of Standards

An important role for standards is to provide scientific and technical support for regulation. It is most desirable to have international standards in place for technologies as they enter commerce so that there can be some harmonization of methods to support trade and some consistency in regulatory requirements. However, the completion and implementation of science-based standards most often occurs after national regulations have been established. Nanotechnology as a field is quite young and few specific national nano-regulatory frameworks exist [42]. In fact, almost every advanced industrialized nation is roughly at the same stage of regulatory assessment as it pertains to nanotechnology; and international coordination, or at least communication, in this area is very strong. The regulatory approach to manufactured nanomaterials is actively debated throughout the world and certain differences in underlying philosophy have already emerged. The U.S. is taking a cautious approach based on risk assessments of nanomaterials. For example, under the Generally Regarded As Safe (GRAS) notification process, adopted by US Food and Drug Administration, a company wanting to use a new food or food-packaging ingredient conducts its own research to determine the ingredient's safety. Other countries and regions have disparate views on requirements, and if not addressed early on, these policy differences potentially may lead to international trade barriers.

It is widely recognized that internationally agreed upon, harmonized standards and approaches for nanotoxicity testing would be of great benefit to regulators across the globe. To support this, it is essential that the international community of experts come together in common forums to debate and create these documents. Since there are many standards development organizations working in this area, at a minimum this requires effective communication and coordination among those organizations with similar scope. Effective communication and cooperation will enable the efficient development of standards to maximize the work of volunteer experts and increase the impact of the output in support of the overall objective of protecting the health and safety of workers, consumers and the environment.

8.4.3 The Future of International Cooperation in Nanotoxicology

International cooperation in nanotoxicity standards development is stimulated by the anticipated scale of the required nanomaterial testing. As predictive mechanism-based biological tests are actively pursued, the preferred approach is still case-by-case descriptive review that could overwhelm even the well-funded national program given the exponential growth in nanomaterial product nomenclature. It is projected that assessing the toxic potential of all existing nanomaterials in the U.S. may take up to 50 years with costs running into billions of dollars [43].

An active role in fostering international harmonization of nanotechnology has been assumed by standards-setting bodies such as ISO (see above); however, other non-governmental initiatives are also attempting to provide international consistency in the risk management and risk assessment of nanotechnology. Often these private initiatives form the basis for work within standards development organizations such as ISO to produce formal guidance documents and standards. One such example is the Dupont-Environmental Defense Fund NanoRisk Framework that is being used as one of the core documents in the an ISO Technical Committee (TC) 229 technical report development activity entitled "Nanotechnologies Nanomaterial Risk Evaluation Framework" (ISO/AWI TR 13121).

We have seen an unusually strong consensus at the international level that a science-based approach to the understanding of potential toxicity of nanotechnology is necessary to inform the regulatory decision-makers. Because of this, nanotechnology presents an opportunity for a different model where international agreement on several issues obtained through a standards-development process is followed by the implementation of national regulations. The benefit to this is clear: internationally consistent environmental and occupational safety and health requirements would allow multi-national companies to employ uniform occupational and environmental health and safety programs globally. The rapid, responsible development and implementation of standards and guidelines for nanotoxicity testing provides a clear path to the scientific underpinning of our national and international regulation.

References

1. National Research Council (U.S.) Committee on the Institutional Means for Assessment of Risks to Public Health: Risk Assessment in the Federal Government: Managing the Process. National Academy Press, Washington, DC (1983)
2. Kuzma, J.: Moving forward responsibly: Oversight for the nanotechnology-biology interface. J. Nanopart. Res. **9**, 165–182 (2007)
3. Vamanu, C.I., Høl, P.J., Allouni, Z.E., Elsayed, S., Gjerdet, N.R.: Formation of potential titanium antigens based on protein binding to titanium dioxide nanoparticles. Environ. Sci. Technol. **39**, 9370–9376 (2005)
4. Kim, H.W., Ahn, E.-K., Jee, B.K., Yoon, H.-K., Lee, K.H., Lim, Y.: Nanoparticulate-induced toxicity and related mechanism in vitro and in vivo. J. Nanopart. Res. **11**, 55–65 (2009)
5. Falck, G.C.M., Lindberg, H.K., Suhonen, S., Vippola, M., Vanhala, E., Catalán, J., Savolainen, K., Norppa, H.: Genotoxic effects of nanosized and fine TiO_2. Hum. Exp. Toxicol. **28**, 339–352 (2009)
6. Okuda-Shimazaki, J., Takaku, S., Kanehira, K., Sonezaki, S., Taniguchi, A.: Effects of titanium dioxide nanoparticle aggregate size on gene expression. Int. J. Mol. Sci. **11**, 2383–2392 (2010)
7. Zook, J.M., MacCuspie, R.I., Locascio, L.E., Elliott, J.E.: Stable nanoparticle aggregates/agglomerates of different sizes and the effect of their sizes on hemolytic cytotoxicity. Nanotoxicology. doi:10.3109/17435390.2010.536615 (2010)
8. Lockman, P.R., Koziara, J.M., Mumper, R.J., Allen, D.D.: Nanoparticle surface charges alter blood-brain barrier integrity and permeability. J. Drug Target. **12**, 635–641 (2004)
9. Grass, R.N., Stark, W.J.: Physico-chemical differences between particle- and molecule-derived toxicity: Can we make inherently safe nanoparticles? Chimia **63**, 38–43 (2009)
10. Limbach, L.K., Wick, P., Manser, P., Grass, R.N., Bruinink, A., Stark, W.: Exposure of engineered nanoparticles to human lung epithelial cells: Influence of chemical composition and catalytic activity on oxidative stress. Environ. Sci. Technol. **41**, 4158 (2007)
11. Hoshino, A., Fujioka, K., Oku, T., Suga, M., Sasaki, Y.F., Ohta, T., Yasuhara, M., Suzuki, K., Yamamoto, K.: Physicochemical properties and cellular toxicity of nanocrystal quantum dots depend on their surface modification. Nano Lett. **4**, 2163–2169 (2004)
12. Hardman, R.: A toxicological review of quantum dots: Toxicity depends on physicochemical and environmental factors. Environ. Health Perspect. **114**, 165–172 (2006)
13. Elder, A., Gelein, R., Silva, V., Feikert, T., Opanashuk, L., Carter, J., Potter, R., Maynard, A., Finkelstein, J., Oberdorster, G.: Translocation of inhaled ultrafine manganese oxide particles to the central nervous system. Environ. Health Perspect. **114**, 1172–1178 (2006)
14. Buzea, C., Pacheco, I.I., Robbie, K.: Nanomaterials and nanoparticles: Sources and toxicity. Biointerphases **2**, MR17–MR71 (2007)
15. Nelsonrees, W.A., Daniels, D.W., Flandermeyer, R.R.: Cross-contamination of cells in culture. Science **212**, 446–452 (1981)
16. Lacroix, M.: Persistent use of "false" cell lines. Int. J. Cancer **122**, 1–4 (2008)
17. Organisation for Economic Co-operation and Development Environment Directorate: Manufactured nanomaterials: Work programme 2006–2008. http://www.olis.oecd.org/olis/2008doc.nsf/LinkTo/NT00000B76/$FILE/JT03240538.PDF (2008)

18. Organisation for Economic Co-operation and Development Environment Directorate: List of manufactured nanomaterials and list of endpoints for phase one of the OECD testing programme. http://www.olis.oecd.org/olis/2008doc.nsf/LinkTo/NT00003282/$FILE/JT03246895.PDF (2008)
19. Organization for Economic Co-operation and Development: Preliminary guidance notes on sample preparation and dosimetry for the safety testing of manufactured nanomaterials. http://www.olis.oecd.org/olis/2010doc.nsf/linkto/ENV-JM-MONO(2010)25 (2010)
20. ISO TC 229 – Nanotechnologies. http://www.iso.org/iso/iso_catalogue/catalogue_tc/catalogue_tc_browse.htm?commid=381983&development=on
21. Kroll, A., Pillukat, M.H., Hahn, D., Schnekenbutger, J.: Current in vitro methods in nanoparticle risk assessment – limitations and challenges. Eur. J. Pharm. Biopharm. **72**, 370–377 (2009)
22. Ferrari, M.: Beyond drug delivery. Nat. Nanotechnol. **3**, 131–132 (2008)
23. National Cancer Institute: Nanotechnology characterization laboratory business plan. http://www.ncl.cancer.gov/ncl_business_plan.pdf (2005)
24. NIST – NCL Joint Assay Protocol PCC-1: Measuring the size of nanoparticles in aqueous media using batch-mode dynamic light scattering. http://www.ncl.cancer.gov/NCL_Method_NIST-NCL_PCC-1.pdf (2007)
25. National Cancer Institute: Nanotechnology characterization laboratory assay cascade. http://www.ncl.cancer.gov/assay_cascade.asp (2009)
26. International Alliance for NanoEHS Harmonization: Characterization of nanomaterial bio-interaction project plan. http://www.nanoehsalliance.org/sections/Projects (2009)
27. International Alliance for NanoEHS Harmonization: Stage 3: In vitro nanoparticle interactions. http://www.nanoehsalliance.org/sections/Projects/Stage3InVitroNanoparticleInteractions (2009)
28. Hackley, V.A., Fritts, M., Kelly, J.F., Patri, A.K., Rawle, A.F.: Informative Bulletin of the Interamerican Metrology System-OAS, Enabling Standards for Nanomaterial Characterization, pp. 24–29. NIST, Gaithersburg (2009). http://www.sim-metrologia.org.br/docs/revista_SIM_ago2009-c.pdf
29. NanoImpact: http://www.nanoimpactnet.eu
30. The European Network on the Health and Environmental Impact of Nanomaterials: Major information package: End of 1st year report. http://www.nanoimpactnet.eu/object_binary/o3043_MIP2_2009-07-07.pdf (2009)
31. Worth, A.P., Balls, M.: The role of ECVAM in promoting the regulatory acceptance of alternative methods in the European Union. Altern. Lab. Anim. **29**(5), 525–535 (2001)
32. NanoInteract: Objectives. http://www.nanointeract.net/sections/AboutNanoInteract/Objectives (2009)
33. Cedervall, T., Lynch, I., Lindman, S., Berggård, T., Thulin, E., Nilsson, H., Linse, S., Dawson, K.A.: Understanding the nanoparticle protein corona using methods to quantify exchange rates and affinities of proteins for nanoparticles. Proc. Natl. Acad. Sci. U. S. A. **104**, 2050–2055 (2007)
34. Jiang, J.K., Oberdorster, G., Biswas, P.: Characterization of size, surface charge, and agglomeration state of nanoparticle dispersions for toxicological studies. J. Nanopart. Res. **11**, 77–89 (2009)
35. Organization for Economic Co-operation and Development: Preliminary review of OECD test guidelines for their applicability to manufactured nanomaterials. http://www.olis.oecd.org/olis/2009doc.nsf/LinkTo/NT000049AE/$FILE/JT03267900.PDF (2009)
36. Sung, J.H., Shuler, M.L.: A micro cell culture analog (mu CCA) with 3-D hydrogel culture of multiple cell lines to assess metabolism-dependent cytotoxicity of anti-cancer drugs. Lab Chip **9**, 1385–1394 (2009)
37. Baudoin, R., Corlu, A., Griscom, L., Legallais, C., Leclerc, E.: Trends in the development of microfluidic cell biochips for in vitro hepatotoxicity. Toxicol. In Vitro **21**, 535–544 (2007)
38. Carraro, A., Hsu, W.M., Kulig, K.M., Cheung, W.S., Miller, M.L., Weinberg, E.J., Swart, E.F., Kaazempur-Mofrad, M., Borenstein, J.T., Vacanti, J.P., Neville, C.: In vitro analysis of a hepatic device with intrinsic microvascular-based channels. Biomed. Microdevices **10**, 795–805 (2008)
39. Huh, D., Matthews, B.D., Mammoto, A., Montoya-Zavala, M., Hsin, H.Y., Ingber, D.E.: Reconstituting organ-level lung functions on a chip. Science **328**, 1662–1668 (2010)

40. European Commission Scientific Committee on Emerging and Newly Identified Health Risks: Opinion on the appropriateness of existing methodologies to assess the potential risks associated with engineered and adventitious products of nanotechnologies. http://www.files.nanobio-raise.org/Downloads/scenihr.pdf (2005)
41. Organisation for Economic Co-operation and Development Environment Directorate: Preliminary review of OECD test guidelines for their applicability to manufactured nanomaterials. http://www.olis.oecd.org/olis/2009doc.nsf/LinkTo/NT000049AE/$FILE/JT03267900.PDF (2009)
42. Roco, M.C.: Coherence and divergence of megatrends in science and engineering. J. Nanopart. Res. **4**, 9–19 (2002)
43. Choi, J.Y., Ramachandran, G., Kandlikar, M.: The impact of toxicity testing costs on nanomaterial regulation. Environ. Sci. Technol. **43**, 3030–3034 (2009)

Chapter 9
Health and Safety Standards*

Vladimir Murashov and John Howard

9.1 Introduction

Health and safety standards aim at minimizing risk to people and the environment. Often, though, there is a significant time lag between the emergence of any new technology and the generation of sufficient risk information to allow a thorough risk assessment and to write a traditional regulatory quantitative risk management standard [1]. In the early twenty-first century, this time lag is leading society to aim to proactively manage the risks of emerging technologies like nanotechnology [2]. Proactive risk management can serve as an initial response to a new technology and later can lead to traditional regulatory standards that are based on lengthy risk assessment data collection. Proactive risk management should include, at a minimum, the following essential features (1) qualitative – as opposed to quantitative – risk assessment; (2) strategies to quickly adapt to accumulating risk information as it develops and to refine any risk management recommendations; (3) recommendations based on a level of precaution that is appropriate to ensure no material impairment of human or environmental health occurs from exposure to the new technology; (4) steps that are equivalent across the spectrum of global emerging technology firms; and (5) robust stakeholder involvement that can lead to widespread voluntary cooperation between firms [2]. These features of proactive risk management are particularly applicable for the development of health and safety standards for the rapidly emerging field of nanotechnology.

Since workers bear the greatest health risk from exposure to any emerging technology, most organizations which develop safety and health standards for nanotechnology have focused their efforts initially on the workplace. The workplace safety

*The findings and conclusions in this report are those of the authors and do not necessarily represent the views of the National Institute for Occupational Safety and Health.

V. Murashov (✉)
National Institute for Occupational Safety and Health, Centers for Disease Control and Prevention, U.S. Department of Health and Human Services, Washington, DC, USA
e-mail: vmurashov@cdc.gov

V. Murashov and J. Howard (eds.), *Nanotechnology Standards*,
Nanostructure Science and Technology, DOI 10.1007/978-1-4419-7853-0_9,

and health standards described in this chapter include voluntary, consensus-type standards adopted by the private sector as well as mandatory, or government regulatory, health-related standards. Occupational safety and health standards usually contain the following elements: (1) occupational exposure limits; (2) hazard communication instructions; (3) standard practices, e.g. safety procedures or reference to codes of conduct; and (4) standard guidance, e.g. industrial hygiene guidance for safe handling of nanomaterials. Additional safety and health related standards are covered in other chapters in this book, for example, in Chaps. 3, 7, and 8 on *Reference Materials, Implication Measurements and Biological Activity Testing*. The following subsections of this chapter describe state-of-the-science for each element of the safety and health standards, highlight standards for nanotechnology currently under development nationally and internationally, and map future directions in standards setting.

9.2 Exposure Limits

Exposure limits have been used for over a century to control exposure to a host of chemical and physical agents. They are most often established to control exposures in working environments and to control ambient contamination in air, food and water. Exposure limits are also used to trigger exposure mitigation measures [3]. In the workplace, occupational exposure limits or OELs serve as benchmarks for assessing and controlling exposures in a worker's breathing zone, for triggering the use of personal protective equipment (PPE) when higher order controls do not reduce airborne concentration levels to sufficiently low levels, and for implementing medical surveillance measures. Historically, most OELs were established to minimize the likelihood of adverse effects occurring from exposure to a potentially hazardous chemical or physical agent over the working life of a worker (Section 6(b)(5) in Ref. [4]). The scientific bases for OELs were determined from the observation of workers exposed to the substance (epidemiology) or from the results of laboratory animal studies (toxicology).

For engineered nanomaterials, it is likely that in the foreseeable future most quantitative risk assessments, including dose-response relationships, will involve the extrapolation of animal data to humans. While human epidemiologic studies are considered the most useful for quantitative risk assessment as a basis for regulatory standards, it is not likely that they will be available for some time [5]. In the meantime, there is an increasing amount of data from acute and sub-chronic toxicology animal studies indicating potential health risks from some engineered nanomaterials [6–9] and a wealth of data on adverse health effects resulting from exposures to incidental nanomaterials [10].

Worldwide, only few OELs for engineered nanomaterials have been established. Examples include amorphous silicon dioxide (SiO_2) [11, 12], carbon black [13] and nanoscale titanium dioxide (TiO_2) [14]. In December, 2010, the US National

Institute for Occupational Safety and Health (NIOSH) published a notice requesting comments on the draft Current Intelligence Bulletin "Occupational Exposure to Carbon Nanotubes and Nanofibers" [15]. The bulletin summarized the adverse respiratory health effects that have been observed in laboratory animal studies with single-walled carbon nanotubes, multi-walled carbon nanotubes and carbon nanofibers and provided recommendations for the safe handling of these materials including an OEL set at 0.007 mg/m³.

In addition to the United States (US) activities, other national efforts to develop OELs for engineered nanomaterials are underway in Germany and United Kingdom (UK). The German Federal Institute for Occupational Safety and Health (BAuA) conducted a risk assessment study on photocopier toner emissions, which are composed of incidental nanoparticles [16]. Using Announcement 910, which was issued by The German Ministry's Committee for Hazardous Substances and which established risk factors for carcinogenic substances [17], BAuA reported the following concentration values for respirable biopersistent toner particles: as of 2008, (1) a tolerable risk of 4 in 1,000 is reached at 0.6 mg/m³; (2) an interim acceptable risk of 4 in 10,000 is reached at 0.06 mg/m³, and, as of 2018; (3) an acceptable risk of 4 in 100,000 is reached at 0.006 mg/m³. The photocopier toner emission study was also used by the Institute for Occupational Safety and Health of the German Social Accident Insurance (IFA) to conclude that in accordance with the German Ministry of Labor and Social Affairs' Technical Rule for Hazardous Substances in the Workplace (TRGS 900) [11] the general dust limit of 3 mg/m³ for the respirable fraction does not apply to the nanoscale particle fraction, but should not be exceeded [18].

In light of the paucity of data on nanomaterial hazard and exposure, IFA recommended benchmark limits to be used for an 8-h work shift. The following limits (expressed as an increase in exposure concentrations over background) have been recommended for monitoring the effectiveness of protective measures in the workplace [18]:

1. For metals, metal oxides and other biopersistent granular nanomaterials with a density of >6,000 kg/m³, a particle number concentration of 20,000 particles/cm³ in the range of measurement between 1 and 100 nm should not be exceeded;
2. For biopersistent granular nanomaterials, with a density below 6,000 kg/m³, a particle number concentration of 40,000 particles/cm³ in the measured range between 1 and 100 nm should not be exceeded (Note: for comparison, it is reported that the air in a normal room can contain 10,000 to 20,000 nanoscale particles/cm³, while these figures can reach 50,000 nanoscale particles/cm³ in wooded area and 100,000 nanoscale particles/cm³ in urban streets [19]);
3. For carbon nanotubes, a provisional fiber concentration of 0.01 fibres/cm³ should not be exceeded, based upon the exposure risk ratio for asbestos [20]; and
4. For nanoscale liquid particles (such as fats, hydrocarbons, siloxanes), the applicable maximum workplace limit or workplace limit values should be employed owing to the absence of effects of solid particles.

These recommended benchmark limits are geared to minimizing exposure in accordance with the state of the art in measurements, and have not been substantiated toxicologically. Even where these recommended benchmark limits are observed, a health risk may still exist for workers. Therefore, they should not be confused with health-based OELs [18].

In the UK, the British Standards Institution (BSI) published a public document, PD 6699-2 "Guide to safe handling and disposal of manufactured nanomaterials" [21], which provides risk guidance for the development, manufacture, and use of engineered nanomaterials. In this document, all nanomaterials are grouped into four hazard categories with assigned benchmark exposure levels (BELs). Similar to the BGIA recommendations, BELs are described as "pragmatic guidance levels only" and are derived from OELs for larger particle forms "on the assumption that the hazard potential of the nanoparticle form is greater than the large particle form." First, there is the "fibrous" category, defined as an insoluble nanomaterial with a high aspect ratio (ratio >3:1 and length >5,000 nm), which is assigned a BEL of 0.01 fibres/cm^3 (one-tenth of the asbestos OEL prescribed in the United States of America (USA) and elsewhere). Second, there is the "CMAR" category, defined as any nanomaterial which is already classified in its larger particle form as a Carcinogenic, Mutagenic, Asthmagenic, or Reproductive toxicant. Nanomaterials in the CMAR category are assigned BELs at one tenth of the mass-based OEL for its larger particle form. Third, there is the "insoluble" category, defined as insoluble or poorly soluble nanomaterials not in the fibrous or CMAR category. Nanoparticles in this category are assigned BELs at one-fifteenth (1/15th) of the mass-based OEL for its larger particle form or 20,000 particles/cm^3. Fourth, there is a "soluble" category, defined as a soluble nanomaterial not in fibrous or CMAR category, which is assigned a BEL at one half of the mass-based OEL for its larger particle form.

In the USA, a programmatic approach based on a national public–private partnership has been proposed for protecting workers from nanomaterials in lieu of mandatory standards. The proposal includes generic provisions for exposure assessment, risk controls, medical surveillance, and worker training [1]. As the quantitative assessment of the nanotechnology risks emerge, the information generated, collected and utilized by the proposed National Nanotechnology Partnership Program [1] could serve as "tentative" OELs [22]. Subsequently, if sufficient evidence of "significant risk" becomes available for a specific nanomaterial, a mandatory occupational health standard could be developed by government. Such a national partnership could help overcome the significant time lag between the generation of sufficient risk assessment information to conduct a thorough quantitative risk assessment and the time needed to write a mandatory governmental regulatory occupational risk management standard. The regulatory requirements in the USA for setting occupational safety and health standards have generally precluded regulators from taking incremental and precautionary steps toward protective standards on the basis of less-than-complete quantitative risk assessment and control information [1].

Worldwide efforts aimed at developing OELs for engineered nanomaterials are intensifying [23]. Those efforts were reviewed at OECD workshops on Exposure Assessment in 2008 [24] and Risk Assessment for Nanomaterials in 2009 [25].

The discussion revealed on-going concerns about the acceptable level of risk, acceptable uncertainty factors and acceptable health end-points. At the June, 2009 meeting of the International Organization Standardization's (ISO) Technical Committee 229 (TC 229) Working Group 3 (WG3), an international group of experts working on the draft Technical Specification "Guide to safe handling and disposal of manufactured nanomaterials" agreed that "[it] will contain guidance for how companies/organizations can make their own decisions regarding Benchmark Exposure Limits, including specific examples for how to develop internal benchmarks as well as citing specific guidelines that can be followed" [26]. Industry-wide and in-house exposure limits have been widely used in the absence of, or in addition to, existing regulatory exposure limits [27]. It requires joint efforts by industry experts in the area of risk assessment and experts on site-specific hazards and exposures familiar with their product and site-specific work environment. Recently, Bayer MaterialScience conducted sub-chronic inhalation studies on MWCNTs and derived in-house an OEL of 0.05 mg/m^3 for its MWCNT product [28]. Nanocyl utilizes a no effect concentration in air of 0.0025 mg/m^3 for an 8-h-per-day exposure [29]. This limit was estimated from the lowest observed adverse effect level of 0.1 mg/m^3 obtained using data from the 90 days inhalation study following OECD 413 test guidelines [8] and by applying an assessment factor of 40 [29].

A number of global efforts are underway to conduct studies aimed at obtaining hazard and exposure data which could be used in quantitative risk assessment analysis to develop OELs. Perhaps the largest effort to generate dose-response and other hazard-related data is OECD Sponsorship Programme for the Testing of Manufactured Nanomaterials [6]. Under this program, OECD member countries, as well as some non-member countries and other stakeholders, are working together to examine the hazard potential of 13 manufactured nanomaterials, which are in, or close to, commercial application [6]. Another Steering Group within the same OECD working party is exploring the feasibility of launching a sponsorship program for exposure assessment for 13 manufactured nanomaterials by conducting a limited number of case studies [30]. The sponsorship program would assemble data that would generate exposure data complementing hazard data for risk assessment analysis [30]. OECD is also looking at a possibility of grouping nanomaterials by hazard potential. Specifically, the Chemicals Committee's Task Force on Hazard Assessment is considering the revision of OECD's guidance on grouping of chemicals [31]. One of the areas under consideration is the possibility to apply the concept of grouping to manufactured nanomaterials, with the aim to fill data gaps by extrapolation or trend analysis [32].

Finally, the World Health Organization (WHO) is the international health organization charged to assist countries to attain "Health for All," and this gives it a unique opportunity to develop solutions for improving safety and health in all countries, especially in developing countries. WHO has the expertise to develop credible and widely accepted approaches in establishing exposure limits [3, 33]. Given the paucity of hazard and exposure data, the WHO could lead the development of guidance on how to establish exposure values in close coordination with OECD efforts.

9.3 Hazard Communication

Hazard communication includes three major categories of information. First, hazard communication includes information passed along the product chain from manufacturers to downstream users and intended to protect workers. Second, hazard communication includes information that accompanies products in transport to warn first-responders and first receivers about specific dangers associated with spills and other accidents. And, third, hazard communication includes information that is designed to inform consumers about specific dangers presented by certain components in consumer products. As a risk management tool, hazard communication is often incorporated into national and international mandatory occupational and environmental standards and plays a large role in product liability laws under a duty to warn of the hazards of a particular product.

9.3.1 Material Safety Data Sheets

Material Safety Data Sheets (MSDSs) provide industrial hygienists, workers, employers and emergency personnel with safety information including guidance about how to safely handle chemical substances. In most countries, manufacturers and importers of chemical substances are required to perform a hazard determination and to report hazard information on MSDS for chemical substances they produce or import (see e.g. [34]). In the USA, the Hazard Communication Standard (29 CFR section 1910.1200) describes the informational elements that are required to be included in a MSDS. Internationally, the Globally Harmonized System for the Labeling and Classification of Chemicals (GHS) was developed to provide a single, harmonized system to classify chemicals, and for producing labels and safety data sheets, with the primary benefit of increasing the quality and consistency of information provided to workers, employers and chemical users. Under the GHS, information on safety data sheets is presented in a designated order.

At this time, however, some authors concluded that MSDSs do not address many characteristics unique to nanomaterials and need to be modified to effectively communicate nanospecific information related to safety and product stewardship [35]. Uncertainty in terminology and nomenclature for nanomaterials also led in some instances to inadequate information being provided on MSDSs [36, 37]. Preparing MSDSs to serve as a source of hazard communication information about a nanomaterial should include at least four important elements (1) a notation about which of the chemical constituents are nano-sized; (2) a notation that the characteristics of nanoparticles may be different from those of the larger particles of the same chemical composition and any data on different properties; (3) a notation that some nanoparticles may initiate catalytic reactions due to their nano size that would not otherwise be anticipated based on their chemical composition alone; and (4) a mechanism to provide updated toxicity information as such information becomes available [38].

Efforts to adjust information contained on MSDSs have been under way in a number of countries led by a range of stakeholders. In Germany, the German Chemical Industry Association (Verband der Chemischen Industrie/VCI) has been developing the "Guidance for the Passing on of Information along the Supply Chain in the Handling of Nanomaterials via Safety Data Sheets" together with stakeholders in dialogue activities [32, 39]. Safe Work Australia is currently in the process of revising the Code of Practice for Safety Data Sheets (SDS) through public consultations [40]. In the section which lists physico-chemical parameters for which information on chemicals should be provided, Safe Work Australia is proposing the addition of a number of non-mandatory parameters, specifically relevant to engineered nanomaterials (but also relevant for some other chemicals):

1. Shape and aspect ratio;
2. Crystallinity;
3. Dustiness;
4. Surface area;
5. Degree of aggregation or agglomeration;
6. Ionisation (redox potential); and
7. Biodurability or biopersistence.

Safe Work Australia is also considering the addition of a small number of advisory notes relating nanotechnologies to other relevant occupational safety and health regulatory documents. For example, the following was added to the draft Policy Proposal for Workplace Chemicals Model Regulations: "*Note: Manufactured nanomaterials may require a different classification and hazard communication elements (labeling and SDS) compared to the macro-form of the same material*" [41].

Internationally, the ISO's TC 229 Work Group 3 is developing a Technical Report on "Preparation of Safety Data Sheets (SDS) for Manufactured Nanomaterials." This effort aims to complement existing MSDS elements described in GHS with nano-specific characteristics predictive of potential health and safety hazards and exposures for engineered nanomaterials.

9.3.2 Labeling

Labeling of regulated substances in consumer products is a risk management tool, which serves to inform consumers about presence of hazardous substances and to allow them make an informed decision on acquiring and using consumer products. In the last 5 years, there have been numerous calls from non-governmental organizations to national governments to institute mandatory labeling of nanomaterials in consumer products especially for nanomaterials in foods and cosmetics [42–48]. It was suggested that such labeling could have ethical and societal benefits by building public trust through transparency and by providing consumers freedom to express their views on broader societal implications of novel technologies [49].

Similar to traditional chemical substances, some nanomaterials can present hazards at certain concentrations and under certain conditions. As with traditional chemical substances, food and cosmetics regulations in most countries provide tools to require producers to disclose the presence of hazardous substances including hazardous nanomaterials. The US Food and Drug Administration (FDA) Task Force on nanotechnology recommended that "the current science does not support a finding that classes of products with nanoscale materials necessarily present greater safety concerns than classes of products without nanoscale materials" [50]. Similarly, according to the opinions of the EU Scientific Committees [Scientific Committee on Emerging and Newly Identified Risks (SCENIHR), on Consumer Products (SCCP) and on food and feed in the European Food Safety Authority (EFSA)] not all nanomaterials induce toxic effects [51]. The Scientific Committees stress that the hypothesis that smaller necessarily means more toxic cannot be substantiated by the published data. However, certain health and environmental hazards have been identified for a variety of manufactured nanomaterials, indicating potential toxic effects. Long, non-degradable, rigid nanotubes (longer than 20 μm) have in several experiments been found to have effects similar to hazardous asbestos, causing inflammatory reactions for instance. Experiments also indicate that carbon nanotubes with these characteristics could induce a specific form of lung cancer, mesothelioma, which is also observed in relation to asbestos exposure. Whether such nanotubes would pose a risk for humans is not known but cannot be ruled out. This means that nanomaterials are similar to other substances, in that some may be toxic and some may not, and some may be toxic only under certain exposure conditions. As there is not yet a generally applicable paradigm for the identification of potential hazards of nanomaterials, the Scientific Committees continue to recommend a case-by-case approach for the risk assessment of nanomaterials [51].

In another example, Food Standards Australia New Zealand (FSANZ) has undertaken a review of its regulatory preparedness in relation to nanotechnology in food including food additives, processing aids, novel foods, contaminants and nutritive substances. As an outcome of this assessment FSANZ has amended its *Application Handbook*, an Australian regulatory instrument, which sets out the essential information required to make an application to vary the *Australia New Zealand Food Standards Code*. The Amendments include the requirement to report particle size, size distribution and morphology where substances are particulate in nature and will remain so in the final food, and where particle size is important to achieving the technological function or may relate to a difference in toxicity. The Amendments do not specifically mention nanomaterials or nanotechnology, but they were introduced to ensure that hazardous nanomaterials and other substances are adequately assessed during the application process [52].

Labeling based on technology or process rather than on a recognized hazard represent a number of challenges related to its usefulness and legitimacy [49]. For instance, such labeling might be inconsistent with national legal frameworks which focus on managing risks associated with specific hazards and would violate the rules of the World Trade Organization (WTO) Technical Barriers to Trade Agreement. Also, labeling poses a danger of information overload. Labeling can confuse rather than inform consumers. In fact, the outcome of such an exercise could be increased risk to consumers because effective hazard communication would be diluted and, in effect, masked [49].

Nevertheless, some countries have adopted nanotechnology specific labeling requirements for consumer products. In 2007, French government launched the Grenelle Project aimed at developing legislation to regulate the manufacture, import or marketing of nanomaterials. The project is organized into two proposed laws: Grenelle 1 and 2. Grenelle 1 is intended to establish general principles, while Grenelle 2 is intended to provide details. Grenelle 1, which was adopted by the French Parliament on July 23, 2009, includes the following requirement relevant to labeling: "The State sets itself the goal that, within 2 years after the law is adopted, the manufacture, importation, or marketing of nanoparticle substances or organisms containing nanoparticles or the product of nanotechnology will become the object of obligatory declaration, notably on quantities and uses, to the administrative authority as well as information to the public and to consumers." Grenelle 2, which was adopted by the French Parliament on August 3, 2009, under Article 73 includes the requirement that "Information related to the identity and uses of these nanoparticle substances shall be publicly available under conditions to be established under the law" [53].

In Russia, the Federal Consumer Rights and Human Well-being Department (*Rospotrebnadzor*) adopted a series of basic regulations covering use of nanomaterials in consumer products including "Regulation 79 regarding the conception of the toxicological studies, risk assessment methodology, methods of identification and quantitative description of nanomaterials." Regulation 79 came into force on October 31, 2007 [54] and states the need for commercial enterprises to inform consumers about the use of nanotechnology products and nanomaterials in consumer products.

The European Parliament adopted regulation in November of 2009 on cosmetic products which requires all producers of cosmetics containing nanomaterials to record their presence on the list of ingredients by using "[nano]" after the names of such ingredients. The scope of reporting on cosmetics products related to nanomaterial is defined as "insoluble or biopersistant and intentionally manufactured material with one or more external dimensions, or an internal structure, on the scale from 1 to 100 nanometers (nm)" [55].

In 2007, BSI released a Publically Available Specification on labeling of nanoparticles and products containing them [56]. The BSI document provides "guidance on the format and content of voluntary labels for manufactured nanoparticles and products or substances containing manufactured nanoparticles…for use by businesses and other organizations involved in the manufacture, distribution, supply, handling, use and disposal of manufactured nanoparticles or products containing manufactured nanoparticles and/or products exhibiting nano-enabled effects." However, until labeling is required by the UK government, the BSI document remains a voluntary guidance. The BSI document also served as an outline for Technical Specification "*Guidance on the labeling of manufactured nanomaterials and products containing manufactured nanomaterials*" under development in the European standardization body, the European Committee for Standardization (CEN), Technical Committee 352 Nanotechnologies. Since the Technical Specification is developed under the Vienna Agreement between ISO and CEN, a limited number of ISO TC 229 experts serve as observers in this CEN activity. The main challenges that this project is facing include: (1) lack of agreed-upon terminology to describe nanomaterials; (2) need to ensure consistency with existing voluntary standards and national and international regulations; (3) need to explain that labeling does not represent judgement about safety or benefits of nanomaterials in the product to avoid consumer confusion at the time of product purchase; and (4) the need to ensure its global rather than regional applicability.

Within the United Nations system, there are food standards developed by the Codex Alimentarius Commission which was created in 1963 by the UN's Food and Agriculture Organization (FAO). The main purposes of food standards are protecting health of the consumers, ensuring fair trade practices in the food trade, and promoting coordination of all food standards work undertaken by international governmental and non-governmental organizations. While the Codex has made progress in a number of areas, an international agreement on standards for the labeling of food products based on emerging technologies, such as biotechnology-aided food products, has so far proved elusive [57]. No activities on nanomaterial labeling for foods have been initiated so far.

9.3.3 Globally Harmonized System

The Globally Harmonized System of Classification and Labeling of Chemicals provides an internationally agreed upon system of hazard classification and labeling and is a common and consistent approach to defining and classifying hazards, and for communicating hazard information on labels and material safety data sheets [58]. The GHS, which is administered within the United Nations system, covers all hazardous chemicals, such as substances, products, and mixtures.

The major target audiences for GHS-based health and safety information include manufacturing workers, consumers, transport workers, and emergency responders and first receivers. Under this system, chemical substances and mixtures are classified according to their physicochemical, health, and environmental hazard characteristics. GHS has been adopted by the European Union and a number of nations. On September 30, 2009 the US Occupational Safety and Health Administration (OSHA) published a proposed rule to align OSHA's Hazard Communication Standard with provisions of the United Nations GHS [59]. Changes to the GHS are made through the Sub-Committee of Experts on the Globally Harmonized System of Classification and Labeling of Chemicals (UNSCEGHS).

Initial discussions on potential modifications to GHS specific to nanomaterials have centered on how the format of Safety Data Sheets can adequately address the novel hazard and exposure potential of nanomaterials. A paper on this matter, prepared by the Australian delegation for the UNSCEGHS meeting in December 2009 [60], proposes that consideration be given to adding the following non-mandatory parameters to *Annex 4 – Guidance on the Preparation of Safety Data Sheets (SDS)*:

1. Particle size and size distribution;
2. Shape and aspect ratio;
3. Crystallinity;
4. Dustiness;
5. Surface area;
6. Degree of aggregation or agglomeration; and
7. Biodurability or biopersistence.

At the December 2009 meeting, it was decided that given the work underway in European Union, OECD, and ISO, the UNSCEGHS will "postpone the consideration of this issue until more information about [nanomaterials] intrinsic properties and characteristics [is] available" [61].

9.4 Risk Mitigation

Standards on nanomaterial risk mitigation have been evolving as more information becomes available on the hazards, exposures and the effectiveness of risk mitigation techniques. Initially, most standards developing organizations focused their efforts on the workplace. For example, the OECD Working Party on Manufactured Nanomaterials (WPMN) Steering Group 8 "Co-operation on Exposure Measurement and Exposure Mitigation" organized its work into three phases (1) exposure in the workplace; (2) exposure to the general population; and (3) exposure to the environment [62, 63].

9.4.1 Occupational Guidance

Within the initial phase covering exposures in the workplace, standardization efforts began with surveys of current practices and general guidance recommending prudent measures to control emissions of nanomaterials in the workplace.

In 2005, one of the first general guidance documents on workplace safety was released by NIOSH as an online internet draft publication called "Approaches to Safe Nanotechnology." After three updates, it was published as a NIOSH numbered publication in 2009 [64]. In regards to exposure mitigation, the document states that according to the current state of the science:

1. For most processes and job tasks, the control of airborne exposure to nanomaterials can be accomplished using a variety of engineering control techniques similar to those used in reducing exposure to general aerosols;

2. The use of good work practices can help to minimize worker exposures to nanomaterials; and
3. Certified respirators provide stated levels of protection [64].

In 2006, the International Council on Nanotechnology's (ICON) "Survey of Current Practices in the Nanotechnology Workplace" [65] was published. The ICON report summarizes results of an international survey of current environmental health and safety and product stewardship practices in the global nanotechnology industry [65]. According to the report:

> Surveyed organizations reported that they believe there are special risks related to the nanomaterials they work with, that they are implementing nano-specific EHS programs and that they are actively seeking additional information on how to best handle nanomaterials. Actual reported EHS practices, however, including selection of engineering controls, PPE, cleanup methods, and waste management, do not significantly depart from conventional safety practices for handling chemicals.…In fact, practices were occasionally described as based upon the properties of the bulk form or the solvent carrier and not specifically on the properties of the nanomaterial.

A number of companies and trade associations have developed safety guidelines for nanomaterials. For example, Degussa (now Evonik) developed voluntary safety and health standards for production facilities working with nanoscale materials [66]. These standards include (1) regular monitoring of microscopic particle concentration in the workplace; (2) health protection of employees through the use of closed systems; and (3) additional technical precautions such as engineering controls and personal protective equipment to maintain concentration of microscopic particles in the air at below 0.5 mg/m^3. In 2007, the German Chemical Industry Association (VCI) and German Federal Institute for Occupational Safety and Health (BAuA) released "Guidance for handling and use of nanomaterials in the workplace" [67]. The VCI/BAuA document provides guidance regarding OSH measures in the production and use of intentionally produced nanomaterials primarily for chemical industry.

In 2008, the OECD WPMN published a survey of national guidance for nanomaterial handling, which highlighted available general industry guidance [68]. In addition, WPMN regularly releases national summaries of activities on safety and health of nanomaterials as Tour-de-Table for WPMN meetings. More specifically for risk mitigation, OECD made public in 2009 its guidance on the use of personal protective equipment [69].

Private standards developing organizations without national membership such as ORC Worldwide and ASTM International also developed guidance available to its members and the public. The ORC website entitled "Nanotechnology Consensus Workplace Safety Guidelines" contains a selection of Health, Safety & Environment tools and reference materials that may be useful to practitioners involved in deployment of nanotechnology [70]. Specifically, there are a number of detailed and practical documents on exposure mitigation on the ORC website (1) General Considerations for Engineering Controls for Nanomaterials (guidance on physical and chemical containment, ventilation and flow extraction, HEPA filtration), (2) Workplace Operational Guidelines (qualitative description of housekeeping standards), and (3) Guidelines for Safe Handling of Nanoparticles in Laboratories (recommendations on exposure risk assessment, engineering controls, PPE and respirators, spill cleanup and disposal). In 2007, ASTM International published "Standard Guide for Handling Unbound Engineered Nanoparticles in Occupational Settings" [71]. This ASTM document describes actions that could be taken in occupational settings to minimize human exposures to unbound, intentionally produced nanometer-scale particles, fibers and other such materials in manufacturing, processing, laboratory and other occupational settings where such materials are expected to be present. It is intended to provide guidance for controlling such exposures as a precautionary measure where relevant exposure standards and/or definitive risk and exposure information do not exist [71].

In 2008, the ISO's TC 229 WG3 "Health, Safety and the Environment" published its first safety and health standard titled "Health and safety practices in occupational settings relevant to nanotechnologies" [72]. The report is based on NIOSH's "Approaches to Safe Nanotechnologies" [64] and aims at assembling the most current information on hazards, exposure assessment and exposure mitigation techniques pertinent to nanotechnologies to facilitate development of site-specific programs by health and safety professionals. Using existing knowledge as a starting point for the control of fine and ultrafine particles (including incidental nanoparticles), guidance is presented for the control of engineered nanomaterials. The Technical Report has become a foundation for the development of national safety and health guidance in a number of countries such as Korea [73], Thailand and Canada. As a next step towards an authoritative normative standard, ISO TC 229 WG3 is developing a Technical Specification "Guide to safe handling and disposal of manufactured nanomaterials" based on the UK BSI guidance with the same title [21].

Mandatory standards on safe handling specific to nanomaterials are implemented in a growing number of countries. Since 2008, US Environmental Protection Agency (USEPA) has been applying its authorities under Section 5(a)(2) describing "Significant New Use Rule" and Section 5(e) describing "Consent Orders" of the Toxic Substances Control Act (TSCA) [74] to require implementation of specific risk mitigation measures for nanomaterials in the workplace including use of NIOSH-approved respirators and wearing gloves and protective clothes. For example, on November 5, 2008 USEPA announced application of Significant New Use Rule (SNUR) to siloxane modified silica and alumina nanoparticles previously registered as P-05-673 and P-05-687, respectively [75]. The generic use of both substances stated in Pre-Manufacture Notices (PMNs) was as an additive. In the ruling EPA announced that "use without impervious gloves or a NIOSH-approved respirator with an [Assigned Protection Factor] of at least ten; the manufacture, process, or use of the substance[s] as a powder; or uses of the substance[s] other than as described in the PMN[s] may cause serious health effects."

On November 6, 2009, USEPA proposed Significant New Use Rules for multi-walled carbon nanotubes and single-walled carbon nanotubes that were the subject of pre-manufacture notices, P-08-177 and P-08-328, respectively [76]. The PMNs describe use of substances as "a property modifier in electronic applications and as a property modifier in polymer composites." According to the notice, these substances are subject to TSCA Section 5(e) consent orders issued by USEPA. The consent orders require protective measures to limit exposures or otherwise mitigate the potential unreasonable risk including wearing a NIOSH-approved full-face respirator with N-100 cartridges, gloves and protective clothing impervious to the chemical substance. The proposed SNURs designate the absence of the protective measures required in the corresponding consent orders as a significant new use.

On February 3, 2010 USEPA proposed SNUR for multi-walled carbon nanotubes, P-08-199, based on determination that "certain changes from the use scenario described in the PMN [Pre-Manufacture Notice] could result in increased exposures" [77]. The PMN states that the substance will be used as an additive/filler for polymer composites and support media for industrial catalysts. In the ruling EPA announced that "use of the substance without the use of gloves and protective clothing, where there is a potential for dermal exposure; use of the substance without a NIOSH-approved full-face respirator with an N100 cartridge, where there is a potential for inhalation exposure; or use other than as described in the PMN, may cause serious health effects."

A Notice issued by the Japanese Ministry of Health, Labor and Welfare (MHLW) to directors of Labour Departments in every prefecture in February 2008 is an example of a specific mandatory governmental general occupational risk management standard for nanomaterials [78]. MHLW revised their Notice in March of 2009 based on recommendations of a committee which was established to discuss safety of nanomaterials in occupational settings [32, 79]. The Notice instructs those involved in the manufacture, repair and inspection of nanomaterials to carry out processes under either sealed, unattended or automated conditions, if there is possibility of exposure to nanomaterials. A local exhaust ventilation system or push–pull type ventilation system must be installed to prevent dispersion of nanomaterials in a location where manufacturing/handling equipment is to be installed which cannot be enclosed or contained. The Notice also instructs to measure concentration of nanomaterials in working environment and provides specific procedures for waste disposal, cleaning, operating procedures, use of protective equipment, health surveillance, worker education etc.

In France, the High Council of Public Health (*Haut Conseil de Santé Publique, HCSP*) issued an Opinion on January 9, 2009 on the safety of workers exposed to carbon nanotubes, in which it recommends mandatory measures. The measures include a requirement that the production of carbon nanotubes, and their use in manufacturing intermediate products and consumer and health products, must be carried out under conditions of strict containment in order to protect workers from aerosolisation and/or dispersion exposure [80]. In addition, through an instruction dated February 18, 2008, the General Directorate for Labour (*Direction Générale du Travail*) reminded its units throughout the country of the legislation governing the prevention of occupational risks arising from exposure to chemical substances containing nanoscale particles. It was emphasized that risk prevention in this field does not lie outside the scope of the regulations of the Labour Code, the provisions of which cover at the very least chemical risk prevention and possibly the special provisions applicable to CMR category 1 and 2 agents (i.e. agents that are carcinogenic, mutagenic or toxic to reproduction) if the substance falls within their scope of application [32].

In 2007, the US Department of Energy (USDOE) published "Approach to Nanomaterials ES&H" [81] to minimize risk to workers in USDOE laboratories. This guidance document formed a basis for a Notice of January 5, 2009, which offered "reasonable guidance for managing the uncertainty associated with nanomaterials whose hazards have not been determined and reducing to an acceptable level the risk of worker injury, worker ill-health and negative environmental impacts" in DOE laboratories [82].

The USDOE Notice provides for safe handling of unbound engineered nanoparticles (UNP) including measures to minimize environmental releases of nanomaterials and requires registries of all nanomaterial workers by requiring establishment of safety and health policies and procedures for activities involving UNP as part of the USDOE-approved Worker Safety and Health Program. [Note: In this document *nanoparticles* are dispersible particles having two or three dimensions greater than 1 nm and smaller than about 100 nm and which may or may not exhibit a size-related intensive property. *Engineered* nanoparticles are intentionally created. This definition excludes biomolecules (proteins, nucleic acids, and carbohydrates), materials for which an occupational exposure limit, national consensus, or regulatory standard exists. Nanoscale forms of radiological materials are also excluded from this definition. *Unbound engineered nanoparticles* are defined by the DOE to mean those engineered nanoparticles that, under reasonably foreseeable conditions encountered in the work, are not contained within a matrix that would be expected to prevent the nanoparticles from being separately mobile and a potential source of exposure.] Specifically, the Notice requires laboratories to:

1. Maintain inventories of nanotechnology activities involving UNP at USDOE sites;
2. Maintain registries of all personnel designated as nanomaterial workers;
3. Provide all nanomaterial workers and their supervisors with training specific to nanotechnology activities;
4. Conduct exposure assessment and establish air monitoring program for UNP based on preliminary exposure assessments;
5. Offer baseline medical evaluations to all nanomaterial workers including general physical exam, pulmonary function test, and general blood work;
6. Control exposures to UNP using a risk-based graded approach;
7. Post signs indicating hazards and exposure mitigation requirements; and
8. Have a documented procedure for managing UNP waste.

In December 2010, OECD announced publication of the Compilation and Comparison of Guidelines related to Exposure to Nanomaterials in Laboratories developed under the leadership of the German delegation to OECD WPMN.

This report revealed that a surprisingly large number of research organizations have developed and made publicly available guidance for safe handling of nanomaterials in laboratories [83].

At the same time, activities are underway to provide guidance for Small- and Medium-size Enterprises (SMEs) through the development of control banding tools and easy to understand communication material targeting workers, management and professionals. In 2008, NIOSH published a brochure for employers, managers and safety and health professionals explaining potential hazards, exposures and effective exposure mitigation tools available for nanomaterials in easy to understand terms [84]. In the UK, the Health and Safety Executive published an Information Note on Nanotechnology in 2004 [85], which gives information on the health and safety issues associated with some aspects of nanotechnology including considerations for monitoring, control measures, and personal protective equipment.

In December of 2008, the Swiss Federal Office for Public Health and the Swiss Federal Office for the Environment published the initial version of the precautionary matrix for synthetic nanomaterials, which will be updated on a regular basis to include new scientific knowledge [Note: "In the context of the precautionary matrix, synthetic nanomaterials are those that comprise nanoparticles or nanorods (abbreviated to NPR in the precautionary matrix) that were specially manufactured for a defined purpose. As a general rule, it is recommended that the precautionary matrix be used for all NPR with at least two dimensions smaller than 500 nm"] [86]. The matrix represents a screening tool based on a control-banding approach to estimate the "nano-specific potential risk" of synthetic nanomaterials and of their applications for workers, consumers and the environment, based on parameters such as stability, reactivity and exposure or emission to the environment of nanomaterials. Risk potential is classified and matched with appropriate measures to protect health and the environment. This risk management tool is provided to the industry to be implemented voluntarily as part of the first phase in a national plan to create regulatory framework conditions for the responsible handling of synthetic nanoparticles.

Also in 2008, an international consortium of stakeholders was created to launch and maintain the *GoodNanoGuide* Project [87]. The *GoodNanoGuide* is based on a wiki software platform, and was described as a "collaboration platform designed to enhance the ability of experts to exchange ideas on how best to handle nanomaterials in an occupational setting. It is meant to be an interactive forum that fills the need for up-to-date information about current good workplace practices and highlights new practices as they develop" [87]. Freely available to the public, the *GoodNanoGuide* guidance on handling of nanomaterials in the workplace is organized in a matrix format. The body of the matrix provides links to specific steps to identify hazard, assess exposure potential and choose controls for given common formulations of nanomaterials (e.g., dry powder, liquid dispersion, solid polymer matrix and non-polymer matrix) and common workplace operations (e.g., material unpacking, synthesis, weighing and measuring, dispersing, mixing, spraying, machining, packing, process equipment cleaning, workplace cleaning, spill cleanup, wastemanagement, reasonably foreseeable emergencies). These common formulations and operations represent the highest potential for exposure. The *GoodNanoGuide* could be particularly valuable to SMEs and to safety and health professionals in low and medium-income countries, who often do not have access to commercial standards.

In March of 2009, the ISO TC 229 WG3 approved a project developing Technical Specification TS 12901-2 "Guidelines for occupational risk management applied to engineered nanomaterials based on a control banding approach." Major challenges facing the project are defining hazards and exposure bands of nanomaterials under the paucity of hazard and exposure data and correlating them with an appropriate and limited number of exposure mitigation bands. Resulting proactive control banding method will be based on the synergy of precautionary and pragmatic approaches and will be significantly different from traditional reactive control banding methods.

Safe Work Australia is an independent statutory agency with primary responsibility to improve occupational health and safety and workers' compensation arrangements across Australia's jurisdictions including six states and two territories. In November of 2009, research commissioned by Safe Work Australia recognized the control banding approach "where similar control measures are used within categories of nanomaterials that have been grouped ('banded') according to their exposure potential and hazardous properties, i.e. grouped according to risk," as "an appropriate method because of the current lack of data available for the risk assessment of individual nanomaterials but there is some understanding of hazards posed by different groups of nanomaterials" [88].

The WHO also has a history of utilizing the control banding approach to providing guidance on how to establish site-specific occupational safety and health program for SMEs in developing countries. Specifically, WHO developed a series of Practical Solutions for the Workplace in the form of toolkits [89]. In collaboration with the UN International Labour Organization (ILO), WHO created the International Chemical Control Toolkit [90]. As a first step in this field, WHO initiated development of WHO Guidelines tentatively titled "Protecting Workers from Potential Risks of Manufactured Nanomaterials." The project aims at providing easy to understand and implement guidance for safe handling of nanomaterials in the workplace targeting SME's and other enterprises with limited access to the most advanced exposure measurement and mitigation technologies and industrial hygiene expertise (http://www.who.int/occupational_health/topics/nanotechnologies/en/).

9.4.2 Environmental and Consumer Guidance

Most of the voluntary and mandatory standards for workplace safety and health described in the previous subsection also include measures to control emissions of nanomaterials into the air or water environments. Thus far there have been few mandatory standards development activities specific to engineered nanomaterials and related to the environment and consumer exposures beyond those initial steps.

OECD Working Party on Manufactured Nanomaterials Steering Group 8 is planning a series of projects aimed at providing guidance on mitigating nanomaterial exposures to the environment and consumers [30].

An example of implemented mandatory standards in the area of environmental or consumer protection includes regulatory actions by USEPA. In 2008, USEPA designated certain nanomaterials "new chemicals" and started issuing consent orders for nanomaterials under TSCA Section 5(e) [91, 92]. The consent orders triggered by PMN review can require specific risk mitigation actions to protect the environment. For example, in September, 2008, USEPA issued consent orders for multi-walled carbon nanotubes and single-walled carbon nanotubes that were the subject of pre-manufacture notices, P-08-177 and P-08-328, respectively [76]. The consent order prohibited any predictable or purposeful release of the PMN substance into the waters of the USA.

USEPA has been also monitoring pesticidal claims made for nanotechnology based products as it would for any other chemical-based products. In the September 21, 2007 Federal Register notice EPA stated that any company marketing a product using silver nanoparticles to kill bacteria must provide scientific evidence that particles do not pose unreasonable environmental risk [93]. On March 7, 2008, an EPA regional office fined ATEN Technology/IOGEAR $208K for "selling unregistered pesticides and making unproven claims about their effectiveness" in the form of a "nanoshield" coating on mouse and keyboard.

In another example, the Review Committee on Basic Research into the Environmental Impact of Nanomaterials, Japanese Ministry of the Environment published the "Guideline for Preventive Environmental Impact from Industrial Nanomaterials (March 2009)" [94]. The document instructs that each company must take suitable action for each circumstance in order to control the environmental release of nanomaterials and describes generally recommended measures.

9.4.3 Comprehensive Risk Management Frameworks

Examples of standards which attempt to provide comprehensive risk assessment and risk management frameworks have also been developed. These standards incorporate guidelines on risk evaluation and mitigation throughout the life of a nano-enabled product.

In 2007, the Environmental Defense Fund and the DuPont Corporation launched the *Nano Risk Framework*, which describes a detailed risk assessment and risk management process for ensuring the safe development of nanoscale materials that can be adapted by different companies and organizations [95]. The framework consists of six distinct action elements:

1. Describe the nanomaterial and its application(s);
2. Profile the lifecycle(s) of the nanomaterial;

3. Evaluate risks associated with its use;
4. Determine risk management strategies;
5. Decide, document, and act; and
6. Review and adapt.

Nano Risk Framework was used as an outline for an ISO TC 229 Technical Report under development, which is presently titled "Nanomaterial Risk Evaluation Process."

Another risk management tool for nanotechnology is CENARIOS® [96] which is the first certifiable risk management and monitoring system specifically adapted to nanotechnologies. The system has been developed by TÜV SÜD (Munich, Germany) and the Innovation Society (St. Gallen, Switzerland) and is already being used in practice. The system uses four individually combinable modules "Risk Estimation and Risk Assessment," "Risk Monitoring," "Issues Management" and "Certification" to integrate the latest findings from science and technology as well as societal, legal and market related factors into risk management.

A recent African and Latin American/Caribbean regional meetings on implementation of the Strategic Approach to International Chemicals Management (Abidjan, Côte D'ivoire, 25–29 January 2010 and Kingston, Jamaica, 8–9 March 2010) adopted resolutions instructing the Open Ended Working Group (OEWG) and International Conference on Chemicals Management (ICCM) 3 to include standards in the form of developments and recommendations related to risk management of nanotechnology [97]. The standards would cover occupational, general public and environmental safety and health throughout nanomaterial life-cycle including nanomaterial waste and would be based on the precautionary approach. On March 2, 2010, the Strategic Approach to International Chemicals Management released for public comments a draft outline of a report focusing on nanotechnologies and manufactured nanomaterials including issues of relevance to developing countries and countries with economies in transition [98]. The report will provide overview of the potential risks to (1) human health, (2) to those who work with them in their production, use and disposal, and (3) to the environment and recommendations on how these could be minimized and managed.

9.5 Codes of Conduct

Another type of standard is based on a code of conduct. Codes of conduct (CoC) standards for nanotechnology aim to address ethical and societal dimensions of developing and commercializing nanotechnology. There have been a number of initiatives in this field within individual organizations, stakeholder groups and governments, mostly in Europe [32]. The CoC put in place by BASF [99, 100] is an example of a code limited to one company. It is a voluntary commitment to guide in a responsible manner the actions of BASF's employees. The Code is based on four principles: (1) protection of employees, customers and business partners; (2) protection of the environment; (3) participation in safety research; and (4) open communication and dialogue.

Another CoC, the Responsible Nano Code, was developed by a non-government multi-stakeholder group in the UK. The Responsible Nano Code provides a framework of best practice for organisations working on the development, manufacture, retail or disposal of products using nanotechnologies. Participating organizations agree to abide by Seven Principles of the Responsible Nano Code:

1. Board Accountability: Each organization shall ensure that accountability for guiding and managing its involvement with nanotechnologies resides with the Board or is delegated to an appropriate senior executive or committee;
2. Stakeholder Involvement: Each organization shall identify its nanotechnology stakeholders, proactively engage with them and be responsive to their views;
3. Worker Health and Safety: Each organization shall ensure high standards of occupational health and safety for its workers handling nano-materials and nano-enabled products. It shall also consider occupational safety and health issues for workers at other stages of the product lifecycle;
4. Public Health, Safety and Environmental Risks: Each organization shall carry out thorough risk assessments and minimize any potential public health, safety or environmental risks relating to its products using nanotechnologies. It shall also consider the public health, safety and environmental risks throughout the product lifecycle;
5. Wider Social, Environmental, Health and Ethical Implications and Impacts: Each organization shall consider and contribute to addressing the wider social, environmental, health and ethical implications and impacts of their involvement with nanotechnologies;
6. Engaging with Business Partners: Each organization shall engage proactively, openly and co-operatively with business partners to encourage and stimulate their adoption of the Code; and
7. Transparency and Disclosure: Each organization shall be open and transparent about its involvement with and management of nanotechnologies and report regularly and clearly on how it implements the Responsible Nano Code [101].

The first example of a CoC specifically aimed at nanotechnology usage in consumer products was published in April 2008 by the Switzerland's Food and Packaging Retailers Association (IG DHS) [102]. The Code contains obligations for IG DHS members regarding personal responsibility, procurement of information and information for consumers. Organizations signing the Code have to consider product safety as a first priority, placing on the market only products that can be judged safe according to the best available evidence. Signatory organizations are also responsible to provide open information to consumers about nanotechnology products, in particular ensuring that "*products described as employing nanotechnologies actually contain components and/or modes of action corresponding to these technologies.*"

In February 2008, the European Commission (EC) adopted the Code of Conduct for Responsible Nanosciences and Nanotechnologies Research. The EC

CoC provides EU Member States, employers, research funders, researchers and more generally all individuals and civil society organisations involved or interested in nanosciences and nanotechnologies research with guidelines favouring a responsible and open approach to nanosciences and nanotechnologies research. The EC CoC is based on a set of general principles:

1. Meaning: Nanosciences and nanotechnologies research should be comprehensible to the public;
2. Sustainability: Nanosciences and nanotechnologies research should be safe, ethical and contribute to sustainable development;
3. Precaution: Nanosciences and nanotechnologies research should be conducted in accordance with the precautionary principle;
4. Inclusiveness: Governance of nanosciences and nanotechnologies research activities should be the principles of openness to all stakeholders;
5. Excellence: Nanosciences and nanotechnologies research should meet the best scientific standards;
6. Innovation: Governance of nanosciences and nanotechnologies research activities should encourage maximum creativity, flexibility and planning ability for innovation and growth; and
7. Accountability: Researchers and research organizations should remain accountable for the social, environmental and human health impacts [103].

The EC intends to regularly monitor and revise its CoC biennially in order to take into account developments in nanosciences and nanotechnologies worldwide and their integration in European society.

9.6 Future Directions

9.6.1 Trends and Outlook

Efforts aimed at development of safety and health standards for nanotechnology are in transition. Early efforts have produced standards that are descriptive in nature. Recently, standards that have been developed reflect a more prescriptive approach. The change in approach arises from that fact that more hazard and risk data are being generated and more risk management techniques are being validated. In addition, the scope of nanotechnology standards is expanding to include not only nanotechnology workers, but also to include environmental exposures to the general public, to consumers, and to the air and water environments. The organizational scope and applicability of nanotechnology standards is expanding from the single organization to collaborations between private sector entities and to involvement by industrial associations. Standards for nanotechnology are beginning to demonstrate regional, national and global levels of involvement.

In most developed countries, well-known occupational, environmental and consumer hazards are covered by mandatory governmental standards. These governmental approaches reflect application-dependent acceptable levels of risk and also incorporate application-dependent uncertainty factors into risk assessment calculations. Many governmental organizations across the world believe that unless emerging technologies like nanotechnology bring about novel types of hazards, or revolutionary types of applications, the governments' existing regulatory regime should suffice or undergo minor modifications [32, 104, 105]. The main challenges to mandatory standards development are in addressing how to best incorporate higher levels of uncertainty in assessing risks that have not yet been well quantified and how existing governmental standards development frameworks can be adapted to protect workers, consumers and the public from nanomaterials whose risks have not fully emerged.

9.6.2 Performance-Based Risk Management Program for Nanotechnology

Based on the efforts to date to develop standards to protect workers, consumers, the general public and the environment from potential adverse impacts of nanotechnology, a performance-based risk management approach may be the best format for the near term.

For a general risk management approach to be successful, metrics to measure the progress towards the use and application of nanotechnology in a safe and responsible manner are needed. Three methods to measure such progress should be considered.

In the first method, single indicators are measured to describe a system. Indicators for the occupational safety and health component of a nanotechnology safety program can be categorized into three groups:

1. Physical indicators, such as exposure measurements and control below benchmark levels;
2. Information/education indicators such as adequacy of MSDS, Standard Operating Procedures (SOPs) and training; and
3. Safety and health indicators such as frequency of injuries and fatalities, sick days, worker compensation claims, reduction in use of Personal Protective Equipment if replaced by measures higher up the hierarchy of controls, productivity level, and exposure accidents (e.g., the Seveso II Directive at [106]).

In the second method, quantitative aggregates of several indicators, or indices, are measured. Indices are expressed as a single score by combining various indicators through a scientifically sound normalization, weighing and aggregation.

In the third method, metrics can be classified into frameworks which present large numbers of indicators in qualitative ways [107]. Frameworks do not aggregate data and therefore values of all indicators can be easily observed.

The three methods have advantages and disadvantages. The first method is the simplest, but does not provide the full account of progress towards occupational safety and health within a comprehensive program. The second and the third method are better suited to comprehensively assess safety and health programs. It is easy to measure progress with the second method and there is a full account of input information with the third method. On the other hand, it can be unclear how to determine weight factors in the second method and there could be difficulties in measuring progress with the third method [107].

Using this approach, a periodic (e.g. annual) assessment of the baseline level for indicators is required in order to identify and recognize effective risk management performance. In the case of the workplace, the existing OSHA Voluntary Protection Programs (VPP) could be adjusted to accommodate novel metrics of success for nanotechnology. The VPP began in 1982 to promote a more cooperative approach between government, labor and management to protect workers and influence employers. VPP is a program to recognize places of employment that have achieved, and are committed to maintaining, superior safety and health performance [108]. The VPP is an example of the third approach utilizing frameworks to measure progress. The progress is measured through two tiers of success: the Star Program and Merit Program. In order to be recognized in the Star Program the participants must achieve certain benchmark values of indicators. For example, a 3-year total case incidence rate and a 3-year days away, restricted, and/or job transfer incidence rate must be below at least 1 of the 3 most recent years of specific industry national averages for nonfatal injuries and illnesses published by the Bureau of Labor Statistics. Specific safety and health management system elements and sub-elements must be implemented. The Merit program recognizes participants that have a good safety and health management system, but they must take additional steps to reach Star quality. The VPP's Star Demonstration Program was created to demonstrate the effectiveness of methods for achieving excellence in safety and health management systems that are potential alternatives to current Star requirements. This program could be considered as a basis for a performance-based risk management program for nanotechnology.

Once the performance-based risk management program is shown to be successful within a single country, it could be implemented in other countries. This could be facilitated by United Nations agencies such as ILO and WHO.

9.6.3 Global Health and Safety Standards Development Coordination

Many national and international standards developing organizations have activities in safety and health standards for nanotechnology and nanomaterials. A concerted effort by all major players is necessary to ensure the most effective and safe development of nanotechnology, as well as any other technology whose risk emergence outstrips the ability to generate quantitative risk information in a timely fashion. For instance, public standards setting bodies could specify mandatory requirements, while the private sector could develop technical standards to satisfy risk assessments and risk management requirements. Under such an effort, a public body such as WHO could be tasked to set maximum exposure limits for specific hazards, while private international standards organizations could set operational and methodological standards for achieving these levels. Similarly, the UN GHS program could define adjustments to the format of hazard communication as necessary and private international standards organizations could develop technical standards on measuring new parameters. A consortium of stakeholders could develop quasi-regulatory standards such as control banding approaches to assess and manage risk of nanomaterials to workers, the public and the environment.

9.7 Conclusion

This chapter has described the current efforts to fashion nanotechnology health and safety standards for workers, consumers, the general public and the environment. It is clear, though, that standards development is in its early stages and non-governmental efforts dominate. While several current mandatory safety and health standards are also applicable to nanomaterials, government efforts are underway to facilitate development of mandatory standards specific to nanomaterials. The absence of sufficient quantitative risk assessment information in animals or in humans limits governments in establishing such mandatory standards at this time. Nevertheless, the call for such standards is growing and it may not be too much longer before governments are forced to answer that call.

References

1. Howard, J., Murashov, V.: National nanotechnolgy partnership to protect workers. J. Nanopart. Res. **11**(7), 1673–1683 (2009)
2. Murashov, V., Howard, J.: Essential features of proactive risk management. Nat. Nanotechnol. **4**(8), 467–470 (2009)
3. World Health Organization: Environmental health criteria document no. 170 assessing human health risks of chemicals: derivation of guidance values for health-based exposure limits. World Health Organization, Geneva. http://www.inchem.org/documents/ehc/ehc/ehc170.htm (1994)
4. Occupational Safety and Health Act of 1970. 29 United States Code §§ 651-678. Rule making. The U.S. National Archives and Records Administration, USA (2000)
5. Schulte, P.A., Schubauer-Berigan, M., Mayweather, C., et al.: Issues in the development of epidemiologic studies of workers exposed to engineered nanoparticles. J. Occup. Environ. Med. **51**, 1–13 (2009)
6. Organization for Economic Cooperation and Development: Working party on manufactured nanomaterials: list of manufactured nanomaterials and list of endpoints for phase one of the oecd testing programme, ENV/JM/MONO(2008)13/REV. Organization for Economic Cooperation and Development, Paris. http://www.olis.oecd.org/olis/2008doc.nsf/LinkTo/NT000034C6/$FILE/JT03248749.PDF (2008). Accessed 1 Feb 2009
7. Shvedova, A.A., Kisin, E.R., Mercer, R., Murray, A.R., Johnson, V.J., Potapovich, A.I., Tyurina, Y.Y., Gorelik, O., Arepalli, S., Schwegler-Berry, D.: Unusual inflammatory and fibrogenic pulmonary responses to single walled carbon nanotubes in mice. Am. J. Physiol. Lung Cell. Mol. Physiol. **289**(5), L698–L708 (2005)

8. Ma-Hock, I., Treumann, S., Strauss, V., Brill, S., Luizi, I., Martiee, M., Wiench, K., Gamer, A., van Ravenzwaay, B., Landsiedel, R.: Inhalation toxicity of multi-wall carbon nanotubes in rats exposed for 3 months. Toxicol. Sci. **112**(2), 468–481 (2009)
9. Seaton, A., Tran, L., Aitken, R., Donaldson, K.: Nanoparticles, human health hazard and regulation. J. R. Soc. Interface **7**(1), S119–S129 (2009)
10. Donaldson, K., Borm, P.: Particle Toxicology. CRC Press, Taylor & Francis Group, Boca Raton, FL (2007)
11. BAuA: Ausschuss für Gefahrstoffe, Technische Regeln für Gefahrstoffe 900 (TRGS 900): Arbeitsplatzgrenzwerte. BAuA, Dortmund. http://www.baua.de/de/Themen-von-A-Z/Gefahrstoffe/TRGS/TRGS-900.html (2009). Accessed 26 June 2009
12. Greim, H.: Gesundheitsschädliche Arbeitsstoffe: Amorphe Kieselsäuren, Toxikologischarbeitsmedizinische Begründung von MAK-Werten. WILEY-VCH, Weinheim (1989)
13. The Japan Society for Occupational Health: Recommendation of occupational exposure limits (2007–2008). J. Occup. Health **49**, 328–344 (2007)
14. National Institute for Occupational Safety and Health: NIOSH current intelligence bulletin: evaluation of health hazard and recommendations for occupational exposure to titanium dioxide (draft). NIOSH, Cincinnati. http://www.cdc.gov/niosh/review/public/TIo2/pdfs/TIO2Draft.pdf (2005)
15. National Institute for Occupational Safety and Health: NIOSH current intelligence bulletin: occupational exposure to carbon nanotubes and nanofibers (draft). NIOSH, Cincinnati.http://www.cdc.gov/niosh/docket/review/docket161A/ (2010)
16. BAuA: Tonerstäube am Arbeitsplatz. BAuA, Dortmund. http://www.baua.de/nn_11598/de/Publikationen/Fachbeitraege/artikel17,xv=vt.pdf (2008)
17. BAuA: Risk figures and exposure-risk relationships in activities involving carcinogenic hazardous substances. BAuA, Dortmund. http://www.baua.de/nn_79754/en/Topics-from-A-to-Z/Hazardous-Substances/TRGS/pdf/Announcement-910.pdf? (2008)
18. BGIA: Criteria for assessment of the effectiveness of protective measures. BGIA. http://www.dguv.de/bgia/en/fac/nanopartikel/beurteilungsmassstaebe/index.jsp (2009)
19. SCENIHR: The appropriateness of existing methodologies to assess the potential risks associated with engineered and adventitious products of nanotechnologies. SCENIHR/002/05. SCENIHR. http://ec.europa.eu/health/ph_risk/committees/04_scenihr/docs/scenihr_o_003b.pdf (2006). Accessed 28 Apr 2010
20. BAuA: zur Exposition-Risiko-Beziehung für Asbest in Bekanntmachung zu Gefahrstoffen 910. BAuA, Dortmund. http://www.baua.de/nn_79040/de/Themen-von-A-Z/Gefahrstoffe/TRGS/pdf/910/910-asbest.pdf (2008)
21. BSI. Guide to safe handling and disposal of manufactured nanomaterials. BSI PD6699-2. BSI, London (2007)
22. McGarity, T.O.: Some thoughts on "deossifying" the rulemaking process. Duke Law J. **41**, 385–1462 (1992)
23. Schulte, P.A., Kuempel, E., Murashov, V., Zumwalde, R., Geraci, C.: Occupational exposure limits for nanomaterials: state of the art. J. Nanopart. Res. 12(6), 1971-1987 (2010).
24. Organization for Economic Cooperation and Development: Report of an OECD workshop on exposure assessment and exposure mitigation: manufactured nanomaterials, ENV/JM/MONO(2009)18. OECD, Paris. https://www.oecd.org/dataoecd/15/25/43290538.pdf (2009)
25. Organization for Economic Cooperation and Development: Report of the Workshop on Risk Assessment of Manufactured Nanomaterials in a Regulatory Context, held on September 16–18 2009, in Washington, DC, United States, ENV/JM/MONO(2010)10. OECD, Paris (2010)
26. International Organization for Standardization: ISO/TC 229 nanotechnologies working group 3 – health safety and the environment, project group 6, "guide to safe handling and disposal of manufactured nanomaterials." Draft report, 9 Jun 2009, Seattle, Washington, United States of America, NANO TC229 WG 3/PG 6 012-2009. ISO, Geneva (2009).
27. McHattie, G.V., Rackham, M., Teasdale, E.L.: The derivation of occupational exposure limits in the pharmaceutical-industry. J. Soc. Occup. Med. **38**(4), 105–108 (1988)

28. Bayer MaterialScience: Occupational exposure limit (OEL) for Baytubes defined by Bayer MaterialScience. http://www.baytubes.com/news_and_services/news_091126_oel.html (2010). Accessed 15 Jan 2010
29. Nanocyl: Responsible care and nanomaterials case study nanocyl. Presentation at European Responsible Care Conference, Prague, 21–23rd Oct 2009. http://www.cefic.be/Files/Downloads/04_Nanocyl.pdf (2009). Accessed 23 Apr 2010
30. Organization for Economic Cooperation and Development: OECD Programme on the Safety of Manufactured Nanomaterials 2009–2012: Operational Plans of the Projects, ENV/JM/MONO(2010)11. OECD, Paris (2010)
31. Organization for Economic Cooperation and Development: Guidance on grouping of chemicals ENV/JM/MONO(2007)28. OECD, Paris. http://www.olis.oecd.org/olis/2007doc.nsf/LinkTo/NT0000426A/$FILE/JT03232745.PDF (2007). Accessed 8 Oct 2009
32. Organization for Economic Cooperation and Development: Current developments in delegations and other international organizations on the safety of manufactured nanomaterials – Tour de Table, ENV/JM/MONO(2009)23. OECD, Paris. http://www.olis.oecd.org/olis/2009doc.nsf/LinkTo/NT000049A2/$FILE/JT03267889.PDF (2009)
33. World Health Organization: Methods used in establishing permissible levels in occupational exposure to harmful agents. Technical Report, No. 601. WHO, Geneva. http://whqlibdoc.who.int/trs/WHO_TRS_601.pdf (1977)
34. Australian National Occupational Health & Safety Commission: National Code of Practice for the Preparation of Material Safety Data Sheets 2nd Edition, NOHSC:2011. NOHSC, Canberra (2003)
35. Conti, J.A., Killpack, K., Gerritzen, G., Huang, L., Mircheva, M., Delmas, M., Harthorn, B.H., Appelbaum, R.P., Holden, P.A.: Health and safety practices in the nanomaterials workplace: results from an International Survey. Environ. Sci. Technol. **42**(9), 3155–3162 (2008)
36. Lam, C.-W., James, J.T., McCluskey, R., Arepalli, S., Hunter, R.L.: A review of carbon nanotube toxicity and assessment of potential occupational and environmental health risks. Crit. Rev. Tox. **36**, 189–217 (2006)
37. Hallock, M.F., Greenley, P., DiBerardinis, L., Kallin, D.: Potential risks of nanomaterials and how to safely handle materials of uncertain toxicity. J. Chem. Health Saf **16**(1), 16–23 (2009)
38. Hodson, L., Crawford, C.: Guidance for preparation of good material safety data sheets (MSDS) for engineered nanoparticles. AIHCe 2009 Nanotechnology Abstracts. Poster Session 404: Nanotechnology. Paper 355. AIHCe, Portland. http://www.aiha.org/education/aihce/archivedabstracts/2009abstracts/Documents/09%20Nanotechnology%20Abstracts.pdf (2009)
39. VCI, Stiftung Risiko-Dialog: Stakeholder-Dialog. Nanomaterials: communication of information along the industry supply chain. VCI, Bischofszell. http://www.vci.de/template_downloads/tmp_VCIInternet/122769Nanoworkshop_industry_supply.pdf?DokNr=122769&p=101 (2008). Accessed 17 Nov 2009
40. http://www.safeworkaustralia.gov.au/swa/HealthSafety/HazardousSubstances/Proposed+Revisions.htm. Accessed 19 Nov 2009
41. Safe Work Australia: Policy proposal for workplace chemicals model regulations. Safe Work Australia, Canberra. http://www.safeworkaustralia.gov.au/NR/rdonlyres/A39E6FD5-1A68-4DFF-89F8-AC393C3FE17C/0/PolicyProposal_chemicals.pdf (2009)
42. Australian Council of Trade Unions: Nanotechnology – why unions are concerned. Fact sheet, Apr 2009. ACTU, Melbourne. http://www.actu.asn.au/Media/Mediareleases/Nanotechposespossible healthandsafetyrisktoworkersandneedsregulation.aspx (2009)
43. European Environmental Bureau: EEB position paper on nanotechnologies and nanomaterials. EEB, Brussels. http://www.eeb.org/publication/2009/090228_EEB_nano_position_paper.pdf (2009). Accessed 21 Jun 2010
44. BfR: BfR consumer conference on nanotechnology in foods, cosmetics and textiles, 20 Nov 2006. BfR, Berlin. http://www.bfr.bund.de/cm/245/bfr_consumer_conference_on_nanotechnology_in_foods_cosmetics_and_textiles.pdf (2006)

45. Natural Resources Defense Council: Nanotechnology's invisible threat: small science, big consequences. NRDC Issue Paper. NRDC, New York. http://www.nrdc.org/health/science/nano/nano.pdf (2007)
46. International Center for Technology Assessment: CTA and friends of the earth challenge FDA to regulate nanoparticles at FDA hearing, 10 Oct 2006. ICTA, Washington, DC. http://www.icta.org/press/release.cfm?news_id=21 (2006)
47. International Center for Technology Assessment: Principles for the oversight of nanotechnologies and nanomaterials. ICTA, Washington, DC. http://nanoaction.org/nanoaction/doc/nano-02-18-08.pdf (2008)
48. Trans Atlantic Consumer Dialogue: Resolution on consumer products containing nanoparticles. TACD, Brussels. http://www.tacd.org/index2.php? option=com_docman&task=doc_view&gid=215&Itemid=40 (Jun 2009)
49. Falkner, R., Breggin, L., Jaspers, N., Pendergrass, J., Porter, R.: Consumer Labeling of Nanomaterials in the EU and US: Convergence or Divergence? Chatham House, London, UK (2009)
50. U.S. Food and Drugs Administration. Nanotechnology: a report of the US Food and Drug Administration Nanotechnology Task Force. FDA, Washington, DC. http://www.fda.gov/nanotechnology/taskforce/report2007.pdf (2007)
51. EC Scientific Committee on Emerging and Newly Identified Health Risks: Risk assessment of products of nanotechnologies. SCENIHR. http://ec.europa.eu/health/ph_risk/committees/04_scenihr/docs/scenihr_o_023.pdf (2009). Accessed 28 Apr 2010
52. Food Standards Australia New Zealand: Application handbook, commonwealth of Australia. Canberra, Australia. http://www.foodstandards.gov.au/_srcfiles/Application%20Handbook%20as%20at%2025%20August%202009.pdf (2009)
53. http://www.legrenelle-environnement.fr/. Accessed on 19 Nov 2009
54. http://www.rospotrebnadzor.ru/documents/postanov/1344/. Accessed 19 Nov 2009
55. European Parliament, Regulation (EC) No 1223/2009 of the European Parliament and of the Council of 30 Nov 2009, on cosmetic products (recast). Off. J. Eur Union. **L 342**, 59–209 (2009). http://www.salute.gov.it/imgs/C_17_pagineAree_1409_listaFile_itemName_15_file.pdf. Accessed 11 Jun 2010
56. BSI: Guidance on the Labeling of Manufactured Nanoparticles and Products Containing Manufactured Nanoparticles. PAS 130:2007. BSI, London, UK (2007)
57. Sand, P.H.: Labeling genetically modified food: the right to know. Rev. Eur. Commun. Int. Environ. Law **15**(2), 185–192 (2006)
58. United Nations: Globally harmonized system of classification and labeling of chemicals. Rev. 2. United Nations, Geneva. http://www.unece.org/trans/danger/publi/ghs/ghs_rev02/02files_e.html (2007)
59. U.S. Federal Register: 30 Sep 2009 (Volume 74, Number 188, pages 50279–50549). http://www.osha.gov/pls/oshaweb/owadisp.show_document?p_table=FEDERAL_REGISTER&p_id=21110
60. United Nations Sub-committee of Experts on the Globally Harmonized System of Classification and Labeling of Chemicals: Hazard communication issues. Additional information on physical and chemical properties for inclusion on the guidance on the preparation of safety data sheets (SDS). Transmitted by the expert from Australia. ST/SG/AC.10/C.4/2009/11. United Nations, Geneva. http://www.unece.org/trans/main/dgdb/dgsubc4/c42009.html (2009). Accessed 5 Jan 2010
61. United Nations Sub-committee of Experts on the Globally Harmonized System of Classification and Labeling of Chemicals: Report of UNSEGHS on its eighteenth session (9–11 Dec 2009). ST/SG/AC.10/C.4/36. United Nations, Geneva. http://www.unece.org/trans/main/dgdb/dgsubc4/c4rep.html (2009). Accessed 5 Jan 2010
62. Murashov, V., Engel, S., Savolainen, K., Fullam, B., Lee, M., Kearns, P.: Occupational safety and health in nanotechnology and organisation for economic co-operation and development. J. Nanopart. Res. **11**(7), 1587–1591 (2009)

63. Organization for Economic Cooperation and Development. Manufactured nanomaterials: roadmap for activities during 2009 and 2010, ENV/JM/MONO(2009)34. OECD, Paris. http://www.olis.oecd.org/olis/2009doc.nsf/LinkTo/NT00004E1A/$FILE/JT03269258.PDF (2009)
64. U.S. National Institute for Occupational Safety and Health: Approaches to safe nanotechnology: managing the health and safety concerns associated with engineered nanomaterials. NIOSH, Cincinnati. http://www.cdc.gov/niosh/docs/2009-125/ (2009)
65. http://cohesion.rice.edu/CentersAndInst/ICON/emplibrary/ICONNanotechSurveyFullReduced.pdf. Accessed 6 Oct 2009
66. http://www.degussa-nano.com/nano/en/sustainability/safeproduction/
67. http://www.baua.de/nn_49456/en/Topics-from-A-to-Z/Hazardous-Substances/Nanotechnology/pdf/guidance.pdf. Accessed 6 Oct 2009
68. Organization for Economic Cooperation and Development: Working party on manufactured nanomaterials: preliminary analysis of exposure measurement and exposure mitigation in occupational settings: manufactured nanomaterials, ENV/JM/MONO(2009)6. OECD, Paris. http://www.oecd.org/dataoecd/36/36/42594202.pdf (2009). Accessed 6 Oct 2009
69. Organization for Economic Cooperation and Development. Comparison of guidance on selection of skin protective equipment and respirators for use in the workplace: manufactured nanomaterials, ENV/JM/MONO(2009)17, 2009. OECD, Paris. https://www.oecd.org/dataoecd/15/56/43289781.pdf (2009). Accessed 6 Jan 2010
70. http://www.orc-dc.com/?q=node/1962. Accessed 6 Oct 2009
71. ASTM International: ASTM E2535-07 Standard Guide for Handling Unbound Engineered Nanoparticles in Occupational Settings. ASTM International, West Conhohocken (2007)
72. International Organization for Standardization: Health and Safety Practices in Occupational Settings Relevant to Nanotechnologies ISO/TR-12885. ISO, Geneva (2008)
73. Korean Standards Agency: KSA6202: Guidance to Safe Handling of Manufactured Nanomaterials in Workplace/Laboratory. KSA, Seoul (2009)
74. The Toxic Substances Control Act (TSCA) of 1976 (1976) 15 United States Code § 2601 et seq. Rule making. The U.S. National Archives and Records Administration, USA
75. U.S. Environmental Protection Agency: Significant new use rules on certain chemical substances. FR 73(215), pp. 65743–65766 (5 Nov 2008). EPA, Washington, DC. http://edocket.access.gpo.gov/2008/pdf/E8-26409.pdf (2008). Accessed 22 Dec 2009
76. U.S. Environmental Protection Agency: Proposed significant new use rules on certain chemical substances. FR 74(214), pp. 57430–57436 (6 Nov 2009). EPA, Washington, DC. http://edocket.access.gpo.gov/2009/E9-26818.htm (2009). Accessed 22 Feb 2010
77. U.S. Environmental Protection Agency: Proposed significant new use rule for multi-walled carbon nanotubes. FR 75(22), pp. 5546–55551 (3 Feb 2010). EPA, Washington, DC (2010)
78. Japanese Ministry of Health, Labor and Welfare: Immediate measures to prevent exposures to manufactured nanomaterials in the workplace (7 Feb 2008). Japanese Ministry of Health, Labor and Welfare, Tokyo. http://wwwhourei.mhlw.go.jp/hourei/doc/tsuchi/200207-a00.pdf (in Japanese) (2008)
79. Japanese Ministry of Health, Labor and Welfare: Notification on precautionary measures for prevention of exposure etc. to nanomaterials (no. 0331013, 31 Mar 2009). Japanese Ministry of Health, Labor and Welfare, Tokyo. http://www.jniosh.go.jp/joho/nano/files/mhlw/Notification_0331013_en.pdf (2009). Accessed 7 Oct 2009
80. http://www.nanonorma.org/ressources/documentation-nanonorma/hcspa20090107-Exp-NanoCarbone.pdf
81. U.S. Department of Energy. Approach to Nanomaterial ES&H. Rev. 3a, May 2008.http://www.er.doe.gov/bes/DOE_NSRC_Approach_to_Nanomaterial_ESH.pdfAccessed 13 Dec 2010
82. U.S. Department of Energy. Notice: the safe handling of unbound engineered nanoparticles. DOE N 456.1. DOE, Washington, DC. http://www.directives.doe.gov/pdfs/doe/doetext/neword/456/n4561.pdf (2009)
83. Organization for Economic Cooperation and Development: Compilation of nanomaterial exposure mitigation guidelines relating to laboratories, ENV/JM/MONO(2010)47. OECD,

Paris. http://www.oecd.org/officialdocuments/displaydocumentpdf?cote=env/jm/mono(2010)47&doclanguage=en
84. U.S. National Institute for Occupational Safety and Health. Safe nanotechnology in the workplace: an introduction for employers, managers, and safety and health professionals. NIOSH Publication No. 2008-112. NIOSH, Cincinnati. http://www.cdc.gov/niosh/docs/2008-112/ (2008). Accessed 18 Nov 2009
85. http://www.hse.gov.uk/pubns/hsin1.pdf. Accessed 6 Oct 2009
86. Höck, J., Hofmann, H., Krug, H., Lorenz, C., Limbach, L., Nowack, B., Riediker, M., Schirmer, K., Som, C., Stark, W., Studer, C., von Götz, N., Wengert, S., Wick, P.: Precautionary Matrix for Synthetic Nanomaterials. Federal Office for Public Health and Federal Office for the Environment. Berne, Switzerland. http://www.bag.admin.ch/themen/chemikalien/00228/00510/05626/index.html?lang=en (2008). Accessed 18 Nov 2009
87. http://goodnanoguide.org
88. Safe Work Australia: Engineered nanomaterials: evidence on the effectiveness of workplace controls to prevent exposure. Commonwealth of Australia, Canberra. http://www.safework-australia.gov.au/NR/rdonlyres/E3C113AC-4363-4533-A128-6D682FDE99E0/0/EffectivenessReport.pdf (2009). Accessed 18 Nov 2009
89. http://www.who.int/occupational_health/activities/practsolutions/en/index.html. Accessed 19 Nov 2009
90. http://www.ilo.org/public/english/protection/safework/ctrlbanding/index.htm. Accessed 19 Nov 2009
91. U.S. Environmental Protection Agency: TSCA inventory status of nanoscale substances – general approach. EPA, Washington, DC. http://www.epa.gov/oppt/nano/nmsp-inventorypaper2008.pdf. (2008). Accessed 22 Dec 2009
92. U.S. Environmental Protection Agency: Toxic Substances Control Act Inventory Status of Carbon Nanotubes. FR 73(212), pp. 64946–64947 (31 October 2008). EPA, Washington, USA. http://www.epa.gov/fedrgstr/EPA-TOX/2008/October/Day-31/t26026.htm (2008). Accessed 22 Dec 2009
93. U.S. Environmental Protection Agency. Pesticide registration; clarification for ion-generating equipment. FR 72(183), pp. 54039–54041 (21 Sep 2007). EPA, Washington, DC. http://www.epa.gov/EPA-PEST/2007/September/Day-21/p18591.htm (2007). Accessed 22 Dec 2009
94. http://unit.aist.go.jp/ripo/ci/nanotech_society/document/090310_moe_eng.pdf. Acccessed 8 Oct 2009
95. http://www.nanoriskframework.org/. Accessed 6 Oct 2009
96. TÜV SÜD: CENARIOS – Certifiable risk management and monitoring system. TÜV SÜD, Munich. http://www.tuev-sued.de/technical_installations/riskmanagement/nanotechnology (2008). Accessed 21 Jun 2010
97. http://www.saicm.org/index.php?content=meeting&mid=90&menuid=&def=1. Accessed 12 Feb 2010
98. Strategic Approach to International Chemicals Management: Report on nanotechnologies and manufactured nanomaterials. SAICM, Geneva. http://www.saicm.org/index.php?menuid=9&pageid=425&submenuheader= (2010). Accessed 17 Mar 2010
99. BASF: Code of conduct nanotechnology. BASF, Ludwigshafen. http://www.basf.com/group/corporate/en/sustainability/dialogue/in-dialogue-with-politics/nanotechnology/code-of-conduct (2010). Accessed 21 Jun 2010
100. BASF: Nanotechnology at BASF. BASF, Ludwigshafen. http://www.basf.com/group/corporate/cn/function/conversions:/publish/content/innovations/events-presentations/nanotechnology/images/dialog.pdf (2008). Accessed 21 Jun 2010
101. http://www.responsiblenanocode.org. Accessed 19 Nov 2009
102. IG DHS: Code of conduct for nanotechnologies. IG DHS, St. Gallen. http://www.innovationsgesellschaft.ch/media/archive2/publikationen/CoC_Nano technologies_english.pdf (2008)
103. Commission of the European Communities: Commission recommendation of 07/02/2008 on a code of conduct for responsible nanosciences and nanotechnologies research. C(2008) 424

final. Commission of the European Communities, Brussels. http://ec.europa.eu/nanotechnology/pdf/nanocode-rec_pe0894c_en.pdf (2008). Accessed 19 Nov 2009
104. Breggin, L., Falkner, R., Jaspers, N., Pendergrass, J., Porter, R.: Securing the Promise of Nanotechnologies. Towards Transatlantic Regulatory Cooperation. Royal Institute of International Affairs, London, UK (2009)
105. Mantovani, E., Procari, A., Robinson, D.K.R., Morrison, M.J., Geertsma, R.E.: Development in Nanotechnology Regulation and Standards – Report of the Observatory Nano. Institute of Nanotechnology, Glasgo, UK (2009)
106. http://ec.europa.eu/environment/seveso/index.htm
107. Ameta, G., Rachuri, S., Fiorentini, X., Mani, M., Fenves, S.J., Lyons, K.W., Sriram, R.D.: Extending the notion of quality from physical metrology to information and sustainability. J. Intell. Manuf. published online on 27 Oct 2009. doi:10.1007/s10845-009-0333-3
108. U.S. Occupational Safety and Health Administration: Voluntary protection program: policy and procedures manual. CSP 03-01-003 (18 Apr 2008). OSHA, Washington, DC. http://www.osha.gov/OshDoc/Directive_pdf/CSP_03-01-003.pdf (2008). Accessed 12 Jan 2009

Chapter 10
Nanotechnology Standards and International Legal Considerations

Chris Bell and Martha Marrapese

10.1 Introduction

Many emerging technologies of the last century have been structured or defined through the standards development process, sometimes either preceded or eventually followed by adaptations in the law.[1] Through standards development and the imposition of legal requirements, innovations can be integrated by industrial and government stakeholders into the societal and economic fabric. Standards can help create the predictable commercial and legal foundations necessary to support sustainable innovation and development. This path is currently available to nanotechnologies.

Within the legal field, international standards have a variety of roles, such as the de facto rule that is recognized by a government agency, the contractual condition that calls for goods and services to conform with applicable standards, or a reference point for intellectual property protection (e.g., through consensus standards defining technical terms). The prevalence of international standards in corporate governance (e.g., commercial paper, legislation, regulations, and decisions by the judiciary) appears to be growing along with the global nature of the commercial economy. Indeed, private standards offer a framework that is parallel to, and preferably consistent with, governance initiatives.

This chapter offers information on the general legal principles and trends that are associated with the use of standards in the law. It is intentionally broad and international in coverage. An exhaustive treatment of the legal relevance of

[1] In the United States, for example, the Occupational Safety and Health Administration's (OSHA) regulation setting safety requirements for grain elevators was almost 10 years in the making following the introduction of voluntary standard on the same topic. Setting Safety Standards, Regulation in the Public and Private Sectors, Ross E. Cheit, UNIVERSITY OF CALIFORNIA PRESS, 1990, University of California Press, E-books collection 1982–2004, Part 2. http://publishing.cdlib.org/ucpressebooks/view?docId=ft8f59p27j&chunk.id=d0e4470&toc.depth=1&toc.id=d0e4449&brand=ucpress.

C. Bell (✉)
Sidley Austin LLP, Washington, DC, USA
e-mail: cbell@sidley.com

V. Murashov and J. Howard (eds.), *Nanotechnology Standards*,
Nanostructure Science and Technology, DOI 10.1007/978-1-4419-7853-0_10,

standards for any particular place or country was not attempted and readers are reminded of the need to seek out a more thorough understanding of the legal requirements for any specific jurisdiction. The chapter is not directed at any particular nanotechnology product, activity, or process. Where it is relevant and possible, the examples provided highlight nanotechnology applications. Given the dynamic nature of technology development, additional examples of nanotechnology-specific applications will undoubtedly emerge. In addition, the topics addressed in the following discussion will probably change with time.

10.2 Standards Are Not Laws

Although standards can have a significant impact on public and private international law in a number of ways, there are distinctions between voluntary standards and enforceable laws. In particular, conformance to international or other standards, no matter how advanced, does not mean that an entity is also in compliance with applicable law. The successful use of a standard is not a substitute for understanding and carrying out the legal obligations established by competent national authorities.

The minimum legal expectation is that standards users are knowledgeable of and in compliance with the applicable law, directives, regulations, and legal decisions in the countries and regions of the world in which they operate. In addition, official recommendations for how to be compliant with the law can issue from individual competent national authorities in the form of guidelines, manuals of decision, and letter rulings. Any of these legal instruments should be evaluated for whether their requirements can be supplemented with the directions contained in a voluntary standard. Such is the case in the realm of product labeling rules that are imposed by regulation for highly regulated items such as food and pesticides, with which compliance is mandatory and the use of supplementary language is restricted.

Most of the time, it is recognized that standards are not intended to satisfy all current legal requirements. More typically, their purpose is to capture an existing norm, satisfy a particular need, or fill a communication gap. For example, though the International Organization for Standardization (ISO) 14000 Environmental Management Standard Series of standards suggest processes to assist companies in identifying and meeting their legal obligations, they offer no guarantee of compliance with the laws of any particular country. However, they can be effective for putting procedures in place to manage an environmental compliance program, as well as serve as a framework for broader "extra legal" goals such as sustainable development. For that reason, many regulatory authorities, including in the EU and North America, view the ISO 14001 EMS standard as a useful compliance assistance tool, even if it is not legally required.[2]

[2] The Emerging Role of Private Social and Environmental International Standards in Economic Globalization. Jason Morrison, Pacific Institute and Naomi Roht-Arriaza, UC Hastings College of Law. International Environmental Law Committee Newsletter. Volume 1, Number 3. Winter/Spring 2006, pp. 10–35.

The standards writing process can also be central in developing common vocabularies that will assist both the private sector in commercial relationships and regulators as they fashion regulatory responses to technical innovations. For example, early standards efforts to build a common vocabulary in nanotechnology are aimed at filling a perceived communication gap. China took the early lead in being first to establish its *United Working Group for Nanomaterials standardization* in December 2003 and in December 2004 China published its first seven national nanotechnology standards, including its first vocabulary standard, GB/19619-2004.[3]

In 2006, the ASTM International Committee E 56 published ASTM E2456 – 06, Standard Terminology Relating to Nanotechnology.[4] This standard defines novel terminology related to nanotechnology developed for broad multi- and interdisciplinary activities. In approximately the same timeframe, the British Standards Institute (BSI) issued seven early-stage terminology documents.[5] Both ASTM and BSI have made these documents publicly available free of charge in recognition of the importance of a common language in establishing the field. Supplementing this earlier work, the ISO Technical Committee (TC) 229 on Nanotechnologies has a Working Group devoted to establishing international consensus definitions for nanotechnology. The body of ISO vocabulary for nanotechnology has been given an 80004 series designation. Three documents have published with more on the way.

As the needs of nanotechnology develop, vocabulary will evolve accordingly. Government regulators appear to be generally aware of these efforts but are under no binding obligation to conform, particularly where regulatory program needs diverge from standardized definitions. In particular, regulators may have to consider factors other than science and technology as they develop terms and definitions related to nanotechnology, such as the ability to exercise jurisdiction over particular kinds of materials and the practicality of enforcement. However, practical complications can arise if regulatory definitions differ from those that are developed through the international standardization process and are in widespread use.

10.3 Standards and Government Decision-Making

Governments typically have authority to adopt standards into law, or write their own standards and make compliance with them mandatory. For example, in the United States, the National Technology Transfer and Advancement Act (NTTAA) (P.L. 104-113) (March 1996) directs federal regulatory agencies to use applicable

[3] http://shop.bsigroup.com/upload/Standards%20&%20Publications/Nanotechnologies/Nano_Presentation.ppt#308,25,Terminology and nomenclature for nanotechnologies.

[4] http://www.astm.org/Standards/E2456.htm.

[5] http://shop.bsigroup.com/en/Browse-By-Subject/Nanotechnology/Terminologies-for-nanotechnologies/.

voluntary consensus standards, except where their use is inconsistent with law or otherwise impractical.[6] This reflects a policy decision to take advantage of the learning derived from the development and implementation of standards, and to discourage regulators from incurring the costs and burdens that could result from attempts to "reinvent the wheel." To provide a specific technical example, the US Environmental Protection Agency's (EPA's) Test Method 24 for evaluating the volatile organic content (VOC) of surface coatings, which is integral to national efforts to reduce smog forming substances that contribute to the formation of harmful ozone, refers to ASTM's measurement methods that are used to determine compliance with the applicable VOC level.[7] As of this writing, a nanotechnology standard has yet to be adopted as U.S. law through the NTTAA process.

Another established example of governments' use of relevant international and regional consensus-based standards is found in the European Community (EC). There, harmonized Community-wide regulations may be supplemented by voluntary, consensus-based, European standards. The EC has made increasing use of standards in support of its policies and legislation as a means of establishing the free circulation of goods in its internal market. The stated objective of the European Committee on Standardization (CEN) is to remove trade barriers for European industry and consumers.[8] Composed of 31 national members and 19 affiliate members, CEN's role is to harmonize all areas of technical standardization for the EU except electrotechnical (under CENELEC), telecommunications (under ETSI), and automotive, aerospace and steel which have special arrangements. In addition, EC Member States are obliged to notify to the Commission, in draft, of proposed technical regulations and to observe a 3 month standstill period before the regulation is made or brought into force under Directive 98/34/EC. Among its other provisions, this Directive also allows the European Commission to make standardisation requests to the European Standards Organisations (ESOs) to develop and adopt European standards in support of European policies and legislation. European standards, even those developed under a Commission mandate, remain voluntary in their use unless specifically incorporated into enforceable law, though conforming with applicable standards might be taken into account by regulatory authorities when considered whether to take enforcement action or the nature and extent of such action.[9]

---

[6] The US National Institute for Standards and Technology (NIST) has identified more than 20,000 citations of standards incorporated by reference in procurement and regulatory documents. An online interactive database available at http://standards.gov/sibr/query/index.cfm demonstrates the extensive use of voluntary standards throughout the U.S. Government. Eleventh Annual Report on Federal Agency Use of Voluntary Consensus Standards and Conformity Assessment, NIST, NISTIR 7503, May 2008, p. 2.

[7] http://www.epa.gov/ttn/emc/methods/method24.html#wtsa.

[8] According to the ISO, this arrangement also poses major challenges to the formal international standardization system, since it calls for a considerable expansion of European standardization activities. An attempt to address this tension was addressed, however, by the establishment of agreements between the international standardizing organizations of ISO and IEC and their European regional counterparts. The joint agreement between ISO and CEN is known as the "Vienna Agreement."

[9] The following link gives access to a *database of mandates*, together with the access to their full text. http://ec.europa.eu/enterprise/policies/european-standards/standardisation-requests/database-mandates/index_en.htm.

Japanese Industrial Standards (JISs) are voluntary national standards, and are established or revised on the basis of a consensus among suppliers, consumers, academia and all related parties. At the same time, under the Industrial Standardization Law, it is stipulated that technical regulations (laws) shall respect JISs where appropriate. Japanese standards experts report that up to 5,000 items in Japanese regulations quote JISs.[10]

10.4 Standards and Intellectual Property

Standardized vocabularies, measurement techniques, and product specifications contribute to the development and common understanding of intellectual property rights. Nanotechnology has been developing as a field in conjunction with a distinguished and particularly unconventional vocabulary (e.g., "Buckeyball", "nanotube," "nanohorns," "qubit," "nanofiber," and "fibril"). In some instances, the meanings of the terms used in nanotechnology muddle conventional usage (e.g., "particle"). In many cases, our ability to consistently and reliably measure and characterize materials in the nanoscale remains a futuristic goal.

Nonetheless, precision in definitions is highly desirable when patenting the rights to any technology. To the extent that even the technical experts cannot agree on how to define the key concepts and components of nanotechnology, it is also difficult to predictably define and establish property rights as well, which can be a barrier to innovation and investment. Patent offices worldwide identify "prior art" as a critical step to distinguish between existing and new intellectual property). For example, the US Patent and Trademark Office has a search category for "nano-art," defined as disclosures related to "nanostructures," a commonly used but not yet well-defined term.

In the absence of good and agreed upon definitions, and the ability to measure and describe the technology in legal instruments, assigning ownership for purposes of intellectual property or contract (i.e., investments, purchases and sales involving materials at the nanoscale) remains less than precise. The lack of uniformity or inability to characterize a material makes it more difficult to delineate ownership interests. In particular, legal practitioners would find the task of identifying the scope of a patent and assessing the potential validity of existing and future claims more complicated. This, in turn, can discourage investors, who are interested in returns associated with the increasing value of defined assets and certainty. Standards development helps to develop key concepts and methodologies which benefit the work of legal practitioners as they attempt to identify and protect intellectual property interests.

[10]International development and voluntary national standardization – Japanese initiative – World Trade and Standardization, 27–28 September, 2001. Berlin, Germany. Akira Aoki, Council member, JISC. www.ifan.org.

10.5 Standards and Corporate Transactions

The corporate structuring (e.g., mergers, acquisitions, divestitures, formation of joint ventures) legal practice is characterized by complex agreements typically reached under difficult time pressures. Attorneys representing clients who are involved in joint ventures, licensing agreements, sales of assets, and similar issues will look to standards to acquire the vocabulary necessary to define terms used in these agreements to delineate their scope and transfer property interests. Today, model codes are often implemented and enforced by professional associations and influence commercial contract specifications. These codes (e.g., building and electric codes) may also be enacted into law in some jurisdictions.

Product quality certification programs based on voluntary standards have a significant role in contract for purposes of international commerce. As a condition of global business relations in the twenty-first century, it is not infrequent for a company to hold itself out as, or require its business partners to be, independently audited and certified to be in conformance with ISO 9000 quality management standards (or related standards).[11]

Product specifications, quality systems, consistent vocabulary, and minimum commercial practices have a powerful effect on decisions to use and invest in nanotechnologies. Even the simplest commercial contracts involving nanotechnology will be clouded to the extent that there is as yet no general agreement on key technical issues, since the parties may not know what they are selling or buying. To the extent that there is uncertainty about how to define and describe emerging technologies, the value placed on investments in or purchases of such technologies may be discounted to reflect the uncertainty and risk.

10.6 Standards and Environment, Health, and Safety Regulation

According to ISO, "the latest and perhaps most diverse landscape of private standards relates to social and environmental aspects, often with associated claims, certification and labeling programmes."[12] ISO has a prominent role in influencing global policy on

[11] The history of the ISO 9000 series dates back to Mil-Q-9858a, a United States military procurement specification established in 1959. Quality system requirements for suppliers were adopted by the US NASA (National Aeronautics and Space Administration) in 1962, and in 1965, NATO (North Atlantic Treaty Organization) accepted specifications for equipment procurement. BS5750, a voluntary standard published by the British Standards Institution (BSI) in 1979, took on quality systems for the manufacturing sector more broadly, and led to the subsequent adoption of ISO 9000 in 1988. The ISO 9000 series is now employed across a variety of types of businesses and is accepted by more than 100 countries. Peter Emerson, History of ISO 9000, http://ezinearticles.com/?History-of-ISO-9000&id=352833.

[12] International Standards and Private Standards, ISO (2010), www.iso.org/iso/private_standards.pdf, p. 7.

environmental, health, and safety (EHS) issues (e.g., the ISO 14000 family of standards dealing with organizational accountability through environmental management systems, auditing, life cycle assessment and environmental marketing claims).

Underscoring the role of environmental protection and sustainable development in nanotechnologies, ISO's Technical Committee (TC) 229 on nanotechnology standards has a dedicated working group to establish technical reports, specifications, and standards in the area of human health and the environment. TC 229 has already published a Technical Report that summarizes "best practices" in the area of occupational health and safety in the context of nanotechnologies.[13] Its public program of work includes documents that will provide recommendations that may be voluntarily adopted in the areas of physical-chemical parameters for toxicology testing, product stewardship and risk management, toxicology screening, control banding, and safety data sheets.

These same areas are the subject of intense and public discussion among regulators in countries around the globe, and several of them are encompassed within the activities currently underway at the Organization for Economic Co-operation and Development's (OECD) Working Group on Manufactured Nanomaterials. Given the multi-stakeholder participation in the work of TC 229, including by governmental representatives, and the coordination between ISO, OECD, the European Commission and other legal entities, it is likely that the standards process can provide useful EHS tools to organizations involved in nanotechnologies that supplement current regulatory programs. The work of ISO might be viewed as an "advance guard," assembling the collective knowledge of experts from around the world and making it available for immediate application while regulators are still attempting to determine what, if any, unique legal requirements are necessary.

10.7 Standards and Consumers

One use of standards in the area of consumer protection is to adopt product specifications and/or product labeling guidance for the purpose of establishing an expected level of information or transparency to meet consumer needs or expectations. Such standards might help to provide consumers with information associated with purchase decisions. At their most advanced stage of development, standards may take the form of voluntary product or professional certification programs, but standards can respond to the needs of the public in less sophisticated forms. Technical product quality or safety standards in particular, when tailored to a specific

[13] ISO/TR 12885:2008 Nanotechnologies – Health and safety practices in occupational settings relevant to nanotechnologies; http://www.iso.org/iso/iso_catalogue/catalogue_tc/catalogue_tc_browse.htm?commid=381983&published=on&includesc=true. Another document, ISO/TR 13121 – Nanotechnologies – Nanomaterial Risk Evaluation, is scheduled for publication in 2010.

material or product application, can provide users and consumers with confidence in the reproducibility and integrity of the product.

In the area of nanotechnologies, there are efforts toward giving a role to standards in helping businesses communicate benefits and risk to the public about their products. ISO TC 229 has a Consumer and Societal Dimensions Task Force as well as a joint project with CEN, the European Committee on Standards, on labeling products containing manufactured nano-objects. Public communications and nanotechnology is and will remain a hot topic area for the foreseeable future. In a 2010 U.S. survey on public attitudes toward medical applications and physical enhancements thatrely on nanotechnology, the researchers at North Carolina State University and Arizona State University found that when those surveyed knew something about nanotechnologies for human enhancement, they were more supportive of it when they were presented with balanced information about its risks and benefits.[14]

Communicating with consumers, particularly at the early stages of a technology's development, is fraught with uncertainty, setting aside the ever-present question of attempting to determine what consumer interests or needs actually are. Notably, point-of-purchase communication is under discussion in the standards community and the regulatory sector as an emerging consideration for nano-objects and products containing nano-objects. An early attempt at a communication document is the BSI PAS 130.[15] The interplay between mandatory business and consumer label requirements and the supplementary guidance that may develop in the form of specifications or standards will need to be carefully considered for a given product type. This suggests a need to develop carefully a path forward with attention to transparency, a reasonable relationship between the intended purposes of standard and its content, broad stakeholder participation and, wherever possible, decision-making based on facts, not speculation. The standards development process for nanotechnology applications appears to be generally in alignment with these guiding

[14] Hiding Risks Can Hurt Public Support For Nanotechnology, Survey Finds. Science Daily. http://www.sciencedaily.com/releases/2010/05/100504095212.htm. May 4, 2010.

In their survey, participants were segmented and given various illustrations and explanations regarding a nanoscale medical device. One set of participants was shown an unrealistic illustration meant to represent a nanoscale medical device. Another group was given the same image together with a "therapeutic" framing statement that described the technology as being able to restore an ill person to full health. A third segment was given the image, along with an "enhancement" framing statement that described the technology as being able to make humans faster, stronger and smarter. Two additional groups of survey participants were given the image, the framing statements, and information about potential health risks. The last set of participants was not given the image, a framing statement or risk information. The survey included 849 participants, with a margin of error of plus or minus 3.3%. At the end of the day, participants were generally accepting of the associated therapeutic advances expected in public health when they were given realistic and more complete information about nanotechnologies. Correspondingly, the less those surveyed knew or were told about nanotechnology, the more skeptical they became.

[15] 2007. Guidance on the labelling of manufactured nanoparticles and products containing manufactured nanoparticles. http://www.bsigroup.com/en/sectorsandservices/Forms/PAS-130/Download-PAS-130/.

principles. Yet it is equally clear that the central decisions balancing the risks and benefits of specific applications of nanotechnology should not be made in the standards setting process.

10.8 Standards and International Trade

Nanotechnologies are expected to be manufactured, processed, and distributed through an integrated economy of global supply chains and financial relationships. Given the global context within which nanotechnologies are already being developed, creating a common vocabulary and consensus technical standards should enhance the ability of commercial and public interests to cooperate on the rational and responsible development of nanotechnologies.

It is commercially valuable that requirements based on meeting the criteria in certain standards find their way into commercial contracts between buyers and sellers who operate in different parts of the world. Legally binding warranties may be backed by a commitment to follow certain standards.

Standard setting can be used as a spur or barrier to competition. As observed by the U.S. Federal Trade Commission (FTC), "the activities of private standard-setting groups are not inherently anticompetitive; indeed they may be substantially procompetitive."[16] By promoting international harmonization of trade through standards, there is an opportunity to reduce the occurrence of non-tariff barriers that result from a patchwork set of rules (e.g., conflicting or different registrations, inspections, certifications, specifications, quality assurance methods, labels). Recognizing this, the World Trade Organization (WTO) Technical Barriers to Trade (TBT) agreement[17] recognizes that requirements that are based on, or consistent with, consensus international standards are not likely to be considered prohibited non-tariff trade barriers. The presumption of acceptable use enhances the influence of standards on international and national law because the TBT agreement effectively encourages countries to rely on or refer to standards in their lawmaking.

According to ISO, the use of international standards in support of public policy and regulation appears to be increasing.[18] The OECD, Asia-Pacific Economic Cooperation (APEC), and the Southern Common Market (MERCOSUR), all encourage the use of international standards as a way of fostering trade within their membership and with the rest of the world. In many regions of the world, good regulatory practices have encouraged the use of performance-based regulation complemented by the voluntary use of standards.

---

[16]Indian Head, Inc. v. Allied Tube & Conduit Corp., No. 81 Civ. 6250 (S.D.N.Y. June 27, 1986), Brief of the United States and the Federal Trade Commission, Amicus Curiae, at 7 (October 24, 1986).

[17]http://www.wto.org/english/tratop_e/tbt_e/tbt_e.htm.

[18]International Standards and Private Standards, ISO (2010), http://www.iso.org/iso/private_standards.pdf, p. 3.

As established by the WTO TBT Agreement, not all standards are equal. Some are sanctioned for use as the basis for international or national rules with legal force and effect, while others are not. The former group is referred to as international standards while the latter category is styled as private standards.

WTO TBT principles for standards development include transparency, openness, impartiality and consensus, effectiveness and relevance, coherence, and addressing the concerns of developing nations.[19] WTO rules distinguish between standards that do and do not follow these principles. Those that do may be utilized as the basis for regulatory measures as "international standards." An example of such standards are the WTO Solid Sawn Wood Packaging Standard or International Standards for Phytosanitary Measure 15 (ISPM 15), which have been approved for implementation by participating WTO countries. The ISPM 15 standard requires of wood packaging to be heat treated or fumigated, if allowed, prior to export.

In an unusual use of the WTO TBT challenge mechanism, in January 2010 the US Environmental Protection Agency (EPA) was asked to re-open the public comment period for a proposal to regulate, under the US Toxic Substances Control Act (TSCA), certain carbon nanotube products. The extension was granted.[20] The requesting party was the European Economic Community's (EEC) WTO TBT Inquiry Point, which is located within the European Commission's Enterprise and Industry Directorate. The request was submitted through the EEC's Inquiry Point counterpart in the United States, which is the National Institute of Standards and Technology (NIST). The EU subsequently submitted comments on the proposed rulemaking, noting the differences between how the EU regulated carbon nanotubes and proposed US approach and the potential for these regulatory distinctions to create different market situations for EU suppliers in the US than in the EU.[21]

Standards need not adhere to the WTO principles to be useful or have legal consequences. "Private standards" can be the basis for commercial standards or certification programs that effectively operate as gates to the marketplace. For example, the systems of lumber grading that exist in countries such as Canada, the United States, and Malaysia illustrate the robust interaction between private standards and trade.[22] These systems assign a lumber grade that serves as a minimum quality control standard for meeting certain building codes requirements that graded; stamped

[19] Decision of the Committee on Principles for the Development of International Standards, Guides and Recommendations with relation to Articles 2, 5 and Annex 3 of the Agreement, Second Triennial Review of the TBT Agreement, http://docsonline.wto.org/DDFDocuments/t/G/.

[20] 75 Fed. Reg, 1024, January 8, 2010.

[21] Comments from the European Union relating to Notification G/TBT/N/USA/499. Proposed significant new use rules on certain chemical substances, January 14, 2010.

[22] In Canada, the National Lumber Grades Authority is responsible for writing, interpreting and maintaining Canadian lumber grading rules and standards (http://www.nlga.org/app/dynarea/view_article/1.html). In the United States, the National Hardwood Lumber Association publishes grading rules for hardwood (http://www.nhla.com/) while the American Lumber Standards Committee publishes Voluntary Product Standard 20 (PS-20) for softwood lumber (http://www.alsc.org/untreated_ps20_mod.htm). See also, Guide on Grading Malaysian Rubberwood (http://www.ehow.com/way_6190491_guideline-grading-malaysian-rubberwood.html).

lumber is used in all wooden buildings. Each piece of lumber is assigned a grade based on its quality, using rules which consider the intended use of the piece, the size of the piece, its characteristics, and in some cases its species. Quality is affected by the number and/or size of characteristics and the way these characteristics affect the strength and appearance of the product. There are corresponding accreditation programs for groups who write and publish grading rules and supervise uniform timber grading. The ALSC Softwood lumber standard is considered the basis for the sale and purchase of virtually all softwood lumber traded in North America. One can easily envision the development of grading systems for construction industry materials impacted by nanotechnology that, in addition to wood, include materials such as concrete, steel, glass, coatings, and fire protection and detection materials.[23]

Ostensibly, the use of a private product material grading system is "voluntary," yet a competitive commercial position may be difficult to establish or maintain in fact without demonstrating conformance to the standard. Often this type of standard is designed to respond to a demand by producers to establish a common technical or commercial platform for their products to facilitate customer understanding or interchangeability. Technology exists today in gemology,[24] in which a diamond's grading system is injected into the stone using high-resolution grayscale photographs on any size diamond, without impacting the quality of the stone. Nanotechnology also is used to create diamonds ("diamondoids") themselves.[25] Synthetic diamonds have existed for many years, but the early versions were not identical to natural ones, and lacked the same strength. Through nanotechnology, synthetic diamonds are chemically identical to natural ones, and since 2007 the Gemological Institute of America has graded their quality.[26]

The degree to which diamond grading standards – or any other standard – can promote market access and achieve market acceptance will vary depending on the state of the law, the commercial sector, and the area in the world in which they operate. It is conceivable that public acceptance of high quality, inexpensive, synthetic diamond jewelry may be slow in some quarters and enthusiastically embraced in others. The beneficial use of "nanodiamondoids" in medicine is less well known, and presents more technological challenges, and the need for government clearances for drugs and devices means these technologies may still be some time in coming.[27] In contrast, the use of single crystal nanowire diamond

[23] Nanotechnology in Construction – one of the top ten answers to world's biggest problems (May 3, 2005) http://www.aggregateresearch.com/article.aspx?ID=6279&archive=1; Nanotechnology and construction report, Nanoforum – European Nanotechnology Gateway (November 2006); http://www.nanoforum.org/nf06~modul~showmore~folder~99999~scid~425~.html?action=longview_publication.

[24] http://www.israelidiamond.co.il/english/News.aspx?boneID=918&objID=6997.

[25] Stanford University, SLAC Public Lecture – Ultimate Atomic Bling: Nanotechnology of Diamonds, May 25, 2010, http://events.stanford.edu/events/238/23829/.

[26] How to tell synthetic diamonds from natural diamonds, http://www.ehow.com/how_4833499_tell-synthetic-diamonds-natural-diamonds.html#ixzz0qpXl1mOO.

[27] Nanotechnology cancer treatment with diamonds (November 7, 2008), http://www.nanowerk.com/spotlight/spotid=8081.php.

and related materials are among the most promising technological advancements emerging in fibre optics and electronics and appear on the verge of commercial reality.[28]

10.9 Standards and Risk Management

As previously noted, government-enacted technical regulations set out requirements often with the aim of protecting public health and safety, and the environment. They may set out the requirements in generic terms (e.g., essential requirements), or in explicit terms, and they may incorporate, by reference or verbatim, the contents of a voluntary standard for all, or some, of the details. These actions by government can make compliance with voluntary standards a part of, or a presumption of, compliance with the law. Government requirements can, in some jurisdictions, serve to reduce companies' legal exposure associated with the misuse or malfunction of a product by setting a minimum performance floor that, if met, might establish that an organization has met its duties to its customers.

Indeed, a strong interplay between standards and regulation is frequently found in the field of product safety. In the United States, common law requires manufacturers to be knowledgeable experts about their products and to test in a manner commensurate with the product's use.[29] Standards can assist in these evaluations particularly with respect to testing and analytical methods. Standards are relevant to litigation in commercial and negligence cases (e.g., product liability and personal injury) by suggesting a standard of care as a yardstick in emerging areas where the legal standard is not as clear against which the conduct of organizations can be measured.

While failure to conform to accepted voluntary standards may be used as evidence that an organization has been negligent, conformance to such standards, however, does not typically provide an absolute defense. At the outset, conformance with a voluntary standard will rarely, if ever, excuse a failure to comply with the law. The rather generic nature of many standards may also mean that they might not fit the specific circumstances of the incident that led to the legal dispute. Further, particularly in a fast-moving area such as nanotechnology, where new techniques, applications and science seem to appear on almost a daily basis, organizations may be faced with more of moving target when making risk management decisions, rather than being able to simply rely on standards (or even just regulatory requirements). In jurisdictions where individual and class action lawsuits by members of the public are viable, it would be prudent for organizations involved in nanotechnologies to adopt a comprehensive and ongoing approach to risk management, rather than rely solely on either legal compliance or conformance with standards, both of which may trail important technical and EHS developments in the field.

[28] Researchers develop new technique simplifying production of high quality diamonds for electronics, October 28, 2008, http://www.azom.com/news.asp?newsID=14307.

[29] Clarence Borel v. Fibreboard Paper Products Corporation, et al. 493 F.2d 1076 (1973).

10.10 When Standards Are More Strict than the Law

Concerns are sometimes expressed when a standard is more stringent than a legal requirement. This concern sometimes arises in the case of product standards. In general terms, product standards provide a description or definition of a product, and are intended to be interpreted similarly by all concerned parties, covering such factors as product composition, construction, dimension, performance, and vocabulary. What happens, however, when a commercial standard is established that is more stringent than what is required by law? For example, if a commercial standard for minimum pesticide residue limits in certain foods is more stringent than regulatory food safety standards, which limit should be used? There is no question as to the regulatory tolerance being the enforceable limit. Beyond this reality, risk management, among other things, might take into account potential liabilities if one does not conform to an applicable consensus standard, the general validity and applicability of such a standard, the credibility of the existing legal requirements, and associated issues such as the views of stakeholders and customers and competitive advantage. This situation can be complicated if there are competing or different potentially applicable standards, or if there are technical differences (e.g., analytical methods, quality control, and verifiability) between the standards and regulatory requirements.

From the standpoint of the law, such technical standards or specifications are seen as voluntary and advisory only and do not supersede or substitute for the legally enacted national laws and regulations that are established by competent national authorities. An infrequent exception may be found where a law or regulation is generally acknowledged to be seriously outdated, and the more modern and stricter voluntary standard is accepted as industry practice. Occupational exposure limits to chemicals established by the American Conference of Government Industrial Hygienists® (ACGIH) have traditionally been required on Material Safety Data Sheets and are notable in this respect for their widespread use by industry in the United States.[30]

10.11 Incorporating Nanotechnology Standards into the Fabric of the Law

The creation of national and international regulations in nanotechnology is affecting decision-making by private industry, public interest organizations, and government. For private industry, international and voluntary standards are an attractive means of establishing a transparent and level commercial playing field.

---

[30]Proposed regulations eliminating this mandatory practice can be found in the September 30, 2009 Federal Register, at http://edocket.access.gpo.gov/2009/pdf/E9-22483.pdf. OSHA proposes on page 50401 to maintain the requirement to list OSHA's mandatory permissible exposure limits (PELs) on the Safety Data Sheets and not the **TLVs©**.

A consistent set of technical standards for nanotechnology across regions is a desirable commercial goal, particularly for a fast-developing and broadly applicable field such as nanotechnology. The failure to reach international understanding on the technical foundations of nanotechnology, such as key definitions and metrology, as well as on accepted health and safety protocols, would only hinder the credible development and expansion of this technology and the benefits it can provide.

It may be possible to achieve international consensus on these issues more readily and rapidly through the standards process than the frequently more ponderous processes of international law. Further, the economic stratification that can result from divergent national command and control strategies, including a "race to the bottom," is a weakness of international law.[31] Widely accepted voluntary standards can, with surprising speed, provide a framework to facilitate the commercial development of an innovative field such as nanotechnology and related adjustments to positive law without significant government intervention. Early "real world" experience implementing these standards can inform the development of positive law on issues of particular interest to the public and governments, such as environmental protection and public health and safety. This approach can also facilitate consistency across various regional and national regulatory regimes by providing a common frame of reference.

Though there have been efforts to coordinate standards setting and governmental regulatory activities, there are significant differences between the two processes. In many regions, government decisions must take into account competing public views, and are subject to scrutiny from the media and other critics. Regulatory agency decisions are frequently subject to legal and political challenges by the regulated community and other stakeholders. Regulators derive their authority from and are accountable under the law, and they are typically charged with protecting the public interest.

Standards development organizations, on the other hand, are generally not accountable to the public in the same way as regulators, nor are their proceedings typically subject to similar levels of external review. In part for these reasons, nanotechnology standards development is proceeding apace. Most prominent standards development organizations, however, have well-developed procedures establishing minimum requirements for participation (including review and resolution of comments), consensus and transparency. However, the voluntary and self-financing nature of the standards setting process typically limits the diversity and number of the active participants.

These fundamental differences in legal and standards setting processes suggests that standards should not lightly be translated into positive law without the more public checks and reviews associated with the proposal and adoption of laws. This is not to suggest that the standards setting process is inherently flawed: bringing together experts from around the world in a relatively transparent and collegial

---

[31] Private Sector and International Standard-Setting: The Challenge for Business and Government, Virginia Haufler, Carnegie Discussion Paper 3, Study Group on the role of the private sector, http://www.Carnegieendowment.Org/Publications/Index.Cfm?Fa=View&Id=220.

consensus process frequently produces concrete and practical results that can be implemented well in advance of what can be the more deliberate and political legal process. Rather, it is to recognize the limitations of standards and the processes used to create them, and to keep those limitations into account when considering their use for legal purposes.

In the case of nanotechnology, ISO's international standards effort includes participants from government, industry, academia and non-governmental groups. Private commercial interests in promoting safe, efficient, and effective products are largely consistent with the public interest. As the field emerges, there is mutual interest among these groups in developing a commonly understood vocabulary. ISO TC 229 has undertaken significant work in this area, and definitions for core terms such as "nanomaterial" are generating a high degree of activity in the international regulatory community.[32]

It is realistic to assume that regulatory agencies will need definitions that may not be identical to those developed by standards organizations. The greater concern is the potential for having inconsistent scopes in the coverage of ISO definitions and regulatory definitions. For example, ISO TC 229 has defined "nanomaterial" in technical and scientific terms. However, it is a conceivable that a regulator might define "nanomaterial" in terms of risk (e.g., basing the definition in part on potential exposure to free nano-objects). The latter approach could have the result that "nanomaterial" might become synonymous with "risky nanomaterial," an outcome that would be as unfortunate as defining "airplane" only in terms of aircraft more likely to crash. Further, a risk-based definition of "nanomaterial" would exclude from consideration nanomaterials that do not pose risks and are largely beneficial. As regulations and vocabulary emerge together, inconsistent definitions could have consequences with the force and effect of law. In anticipation of this, it will be important that regulators and the public anticipate and recognize the overall structure of the ISO nanotechnologies vocabulary, which is intended to reflect common usage, to the greatest extent possible.

ISO's structured vocabulary uses nanomaterials as the broad generic, overarching (and therefore not extremely precise) term to describe nano-objects and nanostructured materials. In addition, the term nanomaterial is intended to encompass nano-objects and nanostructured materials across the entire range of potential technology applications: defense, electronics, food, packaging, paints, medicine, cosmetics, energy, industrial chemical feedstocks, emulsions, articles, etc. With this structure one might recognize the utility of the term "nanomaterial" to the broader community,

---

[32] Interim Policy Statement on Health Canada's Working Definition for Nanomaterials (February 11, 2010), http://www.hc-sc.gc.ca/sr-sr/consult/_2010/nanomater/draft-ebauche-eng.php. New Nano Rule for EU Cosmetics November 27, 2009, http://www.rsc.org/chemistryworld/News/2009/November/27110901.asp. European Parliament approaches Nanomaterials in Electrical and Electronic Equipment with strong Language and a heavy Hand, Nanotechnology Industry Association News, April 27, 2010, http://www.nanotechia.org/news/global?page=2; European Commission urgently demands science-based Definition of Nanomaterials Nanotechnology Industry Association News, March 4, 2010, http://www.nanotechia.org/news/global?page=2.

and encourage the development of terms in regulation that use more precision with a vocabulary that describes the specific area of jurisdiction being addressed, e.g., "nanoscale" (an ISO defined term)[33] silver, nano-object (also defined),[34] nanoscale component, etc. Nanoparticle has also been defined[35] and is available.

Further areas in which private and public interests in standards development could diverge are the degree to which the parties are risk averse and what constitutes an acceptable level of transparency due to proprietary trade secret protection thought to be needed to guard competitive interests. Indeed, the standards setting process is not well-suited to make non-technical policy decisions, such as what constitutes an acceptable level of human health or environmental risk, though standards can contribute to how such levels may be defined, measured and attained. These areas provide examples to illustrate why standards will contribute, but not replace, legal policies that support and regulate the industrialization of nanotechnology. Government mechanisms may need to address different stakeholder interests and therefore can be expected to be informed but, in the end, may diverge from, the needs of the standards setting community.

10.12 Conclusions

The development of standards is, in many respects, welcomed in the area of nanotechnology. The standards process is but one of many forums where regular communication on nanotechnology developments is taking place, needs are being identified, and existing information is being periodically reviewed and updated through a multi-national and multi-stakeholder process. Voluntarily developed standards are also viewed as valuable tools for addressing pressing international environmental and social policy challenges facing nanotechnology that these challenges be identified and address early on. The organizations that are developing and contributing to standards in nanotechnology can provide valuable expert assistance to industry, consumers and governments.

It is most common for standards to typically be a reflection or codification of accepted practices developed over time. In many instances standards necessarily lag behind technological and scientific developments. In the case of nanotechnologies, however, global cooperation and standardization is viewed as important to the integration of nanotechnologies into the economic and social fabric of society. This means that, in the case of nanotechnologies, standardization is playing a leading, not a following, role. The standards process provides a mechanism for global discussion and integration that is somewhat unique in its degree of openness. Membership and/or participation are available to government, commercial interests,

---

[33] ISO TS 27687 (2008).

[34] Ibid.

[35] Ibid.

and non-governmental public interest organizations alike, which promotes a high degree of cooperation among international experts from a diverse range of interests. The robust discourse and balloting process inherent in standards committees strengthens the utility of the work product. Also, once initiated, standards can be developed relatively rapidly in comparison to the typically lengthy processes that characterize the creation of legal requirements.[36] Yet, it is acknowledged that some entities and countries have the resources to participate more fully than others, which places limitations on the degree of consensus achieved.

In contrast to the standards community, which can proceed on a measured basis through expert consensus, the legal sphere is obligated to operate at the forefront of technological innovation and is tasked with managing its implications, frequently on a rapid and ad hoc basis. The range of legal issues that are implicated by the technological advances enabled by nanotechnology, ranging from intellectual property to environmental health and safety to trade law, reflect the breadth and challenge of the process ahead for incorporating nanotechnology into the fabric of the law.

In the case of nanotechnologies, standards and legal sectors are attempting to arrive at solutions at virtually the same time. Standard setting is meeting a perceived need to communicate the new and best information about nanotechnology as it is received by the experts in real time. The law is being continually reexamined for whether it is up to the task of embracing and managing new developments as, or before, they occur. Standards are developed as needed to efficiently commercialize nanotechnologies, in part because they inform legal instruments associated with commercialization. The standards process is a critical transfer point for the exchange of knowledge, know-how, and ideas for fueling the responsible development of next generation manufacturing and societal advancement, and can also provide valuable inputs to legal frameworks that will govern nanotechnology.

[36] Standards are frequently published within 3 years of initiating the work. http://www.rlc.fao.org/en/prioridades/sanidad/normpub.htm. While the process sometimes appears ponderous and slow to those who are in the middle of it, major legislative and regulatory developments can easily take much longer.

Index

V. Murashov and J. Howard (eds.), *Nanotechnology Standards*, Nanostructure Science and Technology, DOI 10.1007/978-1-4419-7853-0,

MIX
Papier aus verantwortungsvollen Quellen
Paper from responsible sources
FSC® C105338
www.fsc.org

If you have any concerns about our products,
you can contact us on
ProductSafety@springernature.com

In case Publisher is established outside the EU,
the EU authorized representative is:
Springer Nature Customer Service Center GmbH
Europaplatz 3, 69115 Heidelberg, Germany

Printed by Libri Plureos GmbH
in Hamburg, Germany